Perfect Numerical and Logical Test Results

Dr Joanna Moutafi and Marianna Moutafi are chartered psychologists with many years of experience designing and analysing psychometric assessments across a variety of industries.

Other titles in the *Perfect* series

Perfect
Numerical and
Logical Test
Results

Joanna and Marianna Moutafi

BOOKS

Published by Random House Books 2010

10 9 8 7 6 5 4 3 2 1

First published in the United Kingdom in 2010 by

Random House Books
Random House, 20 Vauxhall Bridge Road,
London SW1V 2SA

www.rbooks.co.uk

Addresses for companies within The Random House Group Limited can be found at:
www.randomhouse.co.uk/offices.htm

The Random House Group Limited Reg. No. 954009

A CIP catalogue record for this book
is available from the British Library

ISBN 9781847945464

The Random House Group Limited supports The Forest Stewardship
Council (FSC), the leading international forest certification organisation. All our
titles that are printed on Greenpeace approved FSC certified paper carry the FSC logo.
Our paper procurement policy can be found at www.rbooks.co.uk/environment

Mixed Sources
Product group from well-managed
forests and other controlled sources
www.fsc.org Cert no. TT-COC-2139
© 1996 Forest Stewardship Council

Typeset by Palimpsest Book Production Limited, Grangemouth, Stirlingshire
Printed and bound in Great Britain by CPI Bookmarque, Croydon CR0 4TD

Contents

Introduction

Getting a job these days is like going to war. You are competing against too many candidates, fighting to get the job that you want, at a time when everyone has more things to write in their CV than ever before. And you have to go through an increasingly complex process of online applications, interviews, assessment centres, and some type of psychometric test – most likely a numerical or logical reasoning test. If this is the job that you really want, you need to fight every battle of the recruitment process with all that you have. The battle of passing numerical and logical tests is sometimes the one that makes people the most anxious. But it shouldn't be. This is the battle that you can learn the most about, and that you can prepare for the most. It is also the only battle where you can be sure that you will be judged fairly, because there are right and wrong answers – passing it and getting through to the next stage is not up to someone's personal judgement.

So how do you prepare for this battle? The first step is to 'Meet your opponent'. Learn what these tests are, and exactly how and why they are used by recruiters. Learn everything necessary to become familiar with and understand what you will be facing. This will help you to be prepared when the battle comes; it will lower your anxiety of 'facing the unknown' and it will give you a more positive attitude, which will help when taking the test. This is the first step – it is straightforward and easy.

The second step is to physically 'Get ready for your opponent'. There are many things that you can do to help you perform at your best during the battle. Details like learning what to do the week before the test, the night before the test, or even on the day of the test, to be in the best possible physical condition; knowing how to create an optimum working environment when taking an online test from home; and learning to use 'positive psychology' and relaxation techniques before the test can really make a difference. Learn everything that you can do to be ready for battle; the skills are quick to master and painless to put into action.

The next step is the most crucial one, and also the one that you have to work hardest for: 'Master your opponent'. How do you do this? Initially you learn the moves that your opponent is likely to use against you – the types of question that you could face. You then learn how to fight them. You learn your counter-moves – the solution to each type of question – step by step, with as much detail as necessary to really understand how to fight. And then you practise; you go over each move several times, because practice makes perfect. This is the key to your battle – with practice you can really improve your performance. Numerical and logical reasoning tests are not tests that only maths geniuses can do well at. They follow certain rules which you can learn, and the more you practise using the rules, the quicker and more accurate you can become when taking a test.

Once you have mastered your opponent, there is one more thing to do in order to prepare for your battle: simulate a fight and 'Face your opponent'. You have learned the rules of the fight when taking a numerical and logical test. Now it is time to test yourself to see how well you have prepared. This is a chance for you to put into practice what you have learned – can you quickly identify what kind of solution to use for each question and are your calculations correct? This way you can identify any weaknesses and go back over any particular moves, if necessary.

When you have gone through all these steps you will be ready to go into battle.

Step 1: Meet your opponent

Introducing psychometric tests

Psychometric tests are instruments that are used to measure characteristics of a person, such as aptitude, competence, memory, intelligence and personality, in a systematic way. They can be divided into two broad categories: what we call tests of typical performance and tests of maximum performance.

Tests of typical performance try to estimate how a person would usually behave under given circumstances. These tests are essentially questionnaires that explore people's personality, motivation, interests and behaviour patterns. These types of questionnaires are frequently used by organisations in their selection process because they can 'predict' how a person is likely to behave in their job. They can indicate if the person taking the test is extroverted or introverted, reliable or unreliable, if they are conscientious about their job, if they get anxious easily, and if they abide by rules. Employers then use this information to judge whether that person is suitable for a specific role or a specific work environment.

The other type of psychometric assessment, tests of maximum performance, try to estimate the limit of a person's potential, that is the maximum performance that they can achieve. The most common type of maximum performance tests are reasoning tests. Reasoning tests, also called intelligence or ability tests, come in various forms and measure different aspects of a person's ability to reason. The most common reasoning tests measure numerical, verbal, logical, spatial and general reasoning ability. Employers use these tests to get an understanding of how good a job a person can do.

History of reasoning tests

How did employers come up with the idea for reasoning tests? They didn't! The history of reasoning (or intelligence) tests actually dates back to at least 1882, when Sir Francis Galton, Darwin's cousin, established a laboratory in London. He used this to measure people's ability and reaction time in discriminating visual and auditory stimuli. Although intelligence tests have evolved since then, Galton laid the foundations by measuring aspects of people's potential in a systematic way.

If Galton laid the foundations for these tests, then the man who really built them up into what they are today was Binet. In 1905 two French psychologists, Alfred Binet and Theodore Simon, developed the first intelligence test, which was used to identify children who required some form of special education in school – students with lower reasoning ability. It was soon obvious that children who were better at Binet's reasoning test also performed better at school. This brought about the development of numerous reasoning tests and many more experiments looking at what else, including job performance, correlated with the results of these tests. Experiments showed that people who did better at reasoning tests also tended to do better at their job. In time, organisations realised that they could use these tests to select people who have the potential to perform well in specific roles.

Types of reasoning tests

As mentioned above, reasoning tests can have different themes, the most common being numerical and logical reasoning, verbal, spatial and general ability. All of these tests can appear in two forms: they can be timed or untimed.

Timed vs. untimed tests

Timed (or speeded) tests are tests that involve time pressure. The scope of the questions in these tests is limited and the methods you need to use to answer them are clear from the form of the question. Most people could typically find the correct answers to these tests if they were given sufficient time. The tests, however, are concerned with how many questions people can answer correctly in the allotted time. Since the time allowed to solve them is limited, even the most able people may not be able to complete all the questions. Therefore speeded tests are used to differentiate between people's potential, by looking at how quickly, as well as how accurately, they can solve the questions.

Untimed tests are also known as power tests. In contrast to speeded tests, power tests have the luxury of time. They will typically present a smaller number of more complex questions. The methods needed to answer these questions are not obvious, and working out how to answer the question is the difficult part. Once you have determined this, arriving at the correct answer is usually relatively straightforward. In power tests, not everyone is expected to get all the answers, even with unlimited time. In general, people who perform well in timed tests also perform well in untimed tests, because both types of test measure people's reasoning ability. Since both tests can assess the required ability in candidates, organisations typically choose to use timed rather than untimed tests in their selection process for practical purposes: when candidates are invited to sit a test at the employer's site, it is more convenient to have a time limit. The time limit also makes it harder for people to cheat when taking a test online. Apart from these practical advantages in using timed tests, the tests can also give employers an indication of how potential employees perform under pressure. If you have been invited to take an ability test, it is important to check whether the test will have a time limit. If so, this will indicate that the test will be a speeded test, which means you will need to work out the answers as accurately and as quickly as you can.

Numerical reasoning tests

Numerical reasoning tests measure your ability to reason with numbers. These tests require only a very basic level of maths education in order to successfully complete them – they are measuring numerical ability rather than educational achievement. They assess your ability to perform numerical calculations and to extract numerical information from tables and charts, as well as your understanding of such things as numerical transformations, number sequences and the relationships between numbers. Here is an example of a numerical reasoning question:

EXAMPLE QUESTION: If you paid £634.50, including 17.5% VAT, for a television, how much did the television cost before VAT was added?

    a.   £540.00
    b.   £545.50
    c.   £550.00
    d.   £560.50
    e.   £575.00

EXAMPLE SOLUTION: The correct response is option **a**: £540.00. (The explanation of how to solve these types of questions is given in chapter 3.)

Logical reasoning tests

Logical reasoning tests assess your ability to identify patterns within diagrams, measuring your ability to reason in the abstract. These tests, also called abstract or diagrammatic tests, consist of diagrams that follow certain rules which you need to understand and then, usually, apply. They do not contain any numerical or verbal information, and therefore do not require any prior learning in order to solve them. An example of a logical reasoning question is given below.

EXAMPLE QUESTION:

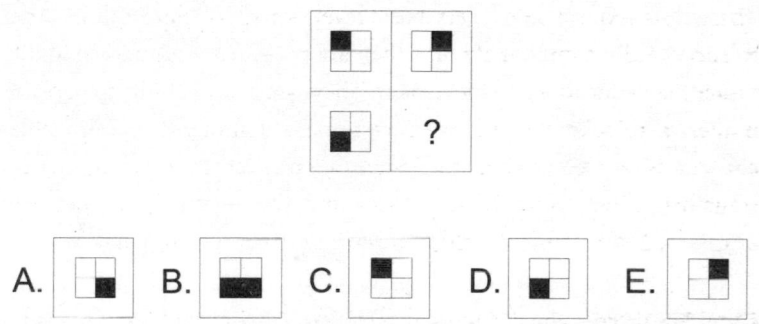

EXAMPLE SOLUTION: The correct answer is option **A**. (The explanation of how to solve these types of questions is given in chapter 4.)

Numerical and logical reasoning tests are some of the most commonly used intelligence tests within occupational settings. This is because they are among the best predictors of job performance. Furthermore, these tests are some of the least likely to discriminate against minority groups and are therefore safer for employers to use.

In this book we will take you through the types of questions that are most frequently asked in numerical and logical tests. We will explain how you can go about solving them and, where possible, we will give you shortcuts that you can use to save time during the test. We will also provide plenty of examples so that you can learn the key rules of these tests and practise them.

Verbal reasoning tests

Verbal reasoning tests measure your ability to reason with verbal information. Examples include simple vocabulary tests or spelling tests, which mostly measure accumulated knowledge rather than reasoning ability. The most typical test of verbal reasoning ability, however, involves passages of text which give you information that you need to interpret. You are then given a list of statements and asked to identify which of the statements is true, based solely on the information provided in the text. Below is an example of such a verbal reasoning item.

EXAMPLE QUESTION: In the battle of Thermopylae of 480 BC an alliance of Greek city-states fought the invading Persian army in a very narrow mountain pass. According to historians, 300 Greek men from Sparta held off a Persian army of millions, in an attempt to buy time for the evacuation of Athens and the preparation of a greater Greek fighting force. Whilst the Spartans inflicted a very high casualty rate on the Persian forces, a traitor, Efialtis, led the Persians through a secret mountain pass to surround them. This led to the death of the 300 soldiers, but gave enough time for the Greeks to prepare and eventually end the Persian invasion at a later battle. Due to the disastrous consequence of Efialtis's betrayal, his name came to be used as the Greek word for 'nightmare'.

Which of the following statements is true, based on the information provided in the passage?

a. The 300 soldiers from Sparta fought the Persian army although it was certain that they would lose.

b. The battle of Thermopylae aided the Greeks in stopping the Persian invasion.

c. The Spartans were able to hold off the Persian army though they were greatly outnumbered, because they fought at a very narrow pass.

d. The Persian invasion could not be stopped due to a traitor who aided the Persians in winning the battle.

EXAMPLE SOLUTION: The correct response is option **b**, as it paraphrases the fourth sentence. The first and third options are neither true nor false, as the text does not provide sufficient information to make this judgement. The last option is false, because the text states that the Persian invasion was stopped at a later battle.

Spatial reasoning tests

Spatial reasoning tests measure your ability to visualise spatial patterns and to mentally manipulate them over a sequence of spatial transformations. These types of tests don't have many rules that you can learn in order to improve your performance. Some people just find it easier than others to mentally manipulate images, but the best way to improve this skill is by practising.

EXAMPLE QUESTION: Which of the following four cubes matches the unfolded cube?

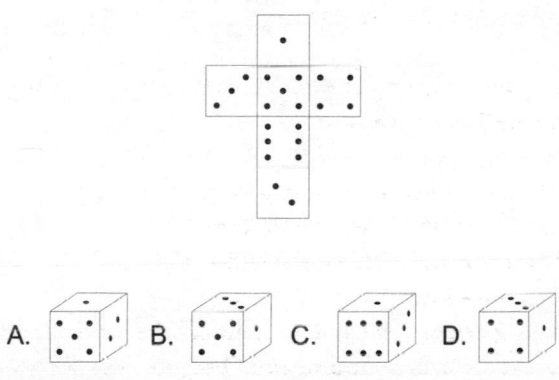

EXAMPLE SOLUTION: The correct answer is option **B**. If you mentally unfold the three sides of the squares that are displayed, and rotate the figure you get anticlockwise, as displayed below, you will see that this matches the original figure.

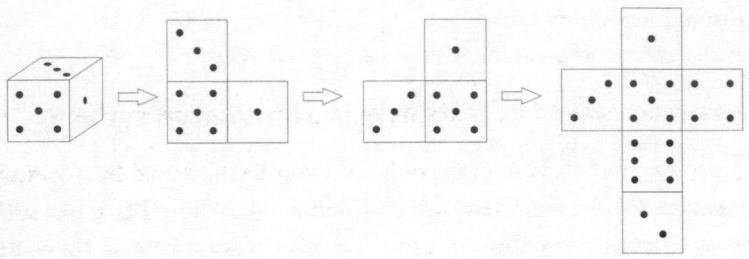

Reasoning tests within recruitment systems

As you are going through the process of 'meeting your opponent', it is useful to have an understanding of how and why reasoning tests are used in the recruitment process of organisations and also get an idea of what other tools may be used.

Why are reasoning tests used?

As we have already mentioned, reasoning tests are used because they can predict how well people will perform the job. There are, of course, a number of external factors that can affect an individual's perform- ance, such as how good their manager is, how clear their goals and responsibilities are, and whether or not they receive constructive feed- back on their performance. These factors, however, are much more dependent on the organisation than the employee, so they are rarely considered during a selection system.

Instead, organisations focus on the personal characteristics that can predict performance and which cannot be easily changed or influenced by the organisation. These include personality, drive and motivation; and, of course, reasoning ability. And reasoning ability can be assessed through reasoning tests.

How much weight do tests have in a recruitment system?

Reasoning tests are rarely the only selection method used by organisations; they are most often used in combination with other tools, such as application forms and interviews. This is because, while they may measure characteristics that are important when it comes to doing a job well, these characteristics are certainly not the only things to affect ability.

So the decision about whether or not to hire you is not usually solely based on reasoning tests. However, the recruitment systems that organisations use are typically divided into stages and you need to pass each stage in order to progress to the next one. This means that even though reasoning tests may be equally weighted with other methods of assessment, when you are at the stage of sitting a test you need to treat the test as if it were the only selection method. Similarly, once you pass it, you need to treat the next stage in exactly the same way.

What other methods of recruitment are used?

Apart from reasoning tests, organisations use tools such as CVs, application forms, questionnaires (such as personality, motivational, integrity, values, culture-fit, and situational judgement questionnaires), interviews, and assessment centres, to assess a candidate's suitability for a given role.

Frequently these tools are used to assess more than one character-istic. Your CV, for example, gives employers an idea of your background. It shows whether you have the required knowledge to do the job from your studies or previous work experience. Your level of education and the grades you've achieved show how 'good' your knowledge is and how much effort you are willing to put in. And characteristics such as persist-ence may be shown by the length of time you've remained in previous jobs. Other methods, such as interviews, can be used to assess your personality, communication skills, motivation, aspirations and whether or not you fit in with the culture of the organisation. Assessment centres can include exercises that simulate the role for which you are applying. They evaluate skills such as teamwork, leadership, presentation and problem-solving, and can also assess your efficiency when working under pressure.

Depending on the organisation's budget for recruitment and on the ratio of applicants to openings, they will select one or more of these methods of assessment. Reasoning tests are increasingly often among the methods selected.

Which types of role are reasoning tests used for?

Reasoning tests could be used for any type of role or position, as long as reasoning ability is an important attribute to have in order to perform well in that particular role. How is this decided? The employer will typically first investigate which characteristics are required for the job, through a process called job analysis.

Job analysis usually involves examining the job specifications of the role and interviewing people who are currently in the role or have previously been in it, and their managers or subordinates, even shad-owing a person who is doing that job. All of these procedures aim to identify the set of characteristics needed for someone to do the job well. Once the required characteristics are identified, the next step is to find what tools can be used to assess them, and to select the most suitable ones, given the role, culture and budget. Once the tools are selected, employers decide what evaluation, or score, they expect their candidates to achieve in order to get the job (or to pass to the next stage of the selection system).

So when reasoning ability is identified as a main characteristic of performing well in a particular role, the next step is to decide how it will be assessed. This means deciding which type of reasoning test (numerical or logical) is most relevant, whether it will be timed or untimed, whether it will be taken on the organisation's premises – and if so, in paper-and-pencil format or on a computer – or whether it will be taken at home, by accessing an online testing website.

How many people pass?

The number of people that pass reasoning tests will vary depending on, among other things, the number of people the organisation needs to recruit and the number of applications they have. It is reassuring, however, to know that organisations don't use reasoning tests in order to select the top, say, 10% of their applicants. Instead, they most typically use them to select around 70% of their applicants. This means that usually only about 30% are deselected through these tests.

The percentage of people who are deselected is relatively low because, as we discussed earlier, reasoning tests can only predict how good people will be at their job to a certain extent. Organisations know that the key point is not that people with the highest reasoning ability will be the best for the job, but that people with a lower reasoning ability will struggle more to succeed. So they deselect people with a lower reasoning ability before assessing whether the remaining candidates possess other characteristics relevant to the job.

What score do you need to pass?

Knowing that only about 30% of the applicants will be deselected does not, however, mean that you only need to respond to about 31% of the questions in order to pass. It means that you need to score higher than about 30% of the other applicants. What does this mean in terms of a score? That depends on how difficult the test is and the ability of the other applicants. If you are applying for a position where numerical calculations will be a big part of the job, the chances are that applicants for this position will be relatively competent with numbers. This means that you will need a better score to pass.

Numerical tests, however, are sometimes also used for jobs that don't require much reasoning with numbers. Interestingly, this is because numerical reasoning ability can actually reflect how well you reason with other things. The good thing is that if you are applying for a position that does not require much reasoning ability, the score needed to pass will be lower. How much lower you can't really know. But whatever the pass mark is, you should always strive to give your best performance.

Should your genius friend take the test for you?

If this question never crossed your mind you can move on to the next section. It does cross the mind of some applicants, however, because the stakes are high. You may be applying for your 'dream job', and think that breaking the rules just this once is worth it, if the ends justify the means.

We won't get into a philosophical argument here. After all, philosophical arguments should be two-way discussions and not monologues! But we will try to answer whether it is safe to ask someone else to take the test for you. This of course could only be done if you were asked to sit the test online. What you need to know is that organisations do consider the fact that when they ask their candidates to sit a test remotely (online), it is possible for candidates to cheat. Organisations are naturally trying to counteract this.

How do they do it? The simplest way is to ask applicants to re-sit the test in their offices. It is, of course, costly for organisations to ask everyone to re-sit the test, so they frequently select a random sample of applicants to go through this procedure. Other companies would rather be safe than sorry and ask everyone to complete it again. And some companies don't think the percentage of applicants who cheat is high enough for them to go to the trouble of identifying them. The point is, you can't know what approach each company is taking – whether or not they will be re-testing applicants who have completed the tests online. So the question is this: should you risk asking someone else to take the test for you, knowing that there is a chance you may then have to take the test again yourself under supervised conditions?

Our advice is that you shouldn't. The potential gain is not high enough to justify losing your credibility. After all, if the organisation decides that you are not suitable for the job as a result of a reasoning test, it is quite likely that the job will not be suitable for you either.

Fairness of reasoning tests

You are quite justified in wondering at this point how fair these tests really are. After all, someone has designed a test that will, to some extent, affect your future. Of course this stands for many other tests, such as A levels and university exams. However, you know what you need to study in order to pass your A levels. But what do you study in order to pass a reasoning test? The answer is not about studying; it is really all about practising. This is because reasoning tests do not typically measure knowledge that you have gathered or facts you have learned; they measure your ability to reason. But it is important for you to know whether reasoning tests are fair, because you will sit the test with a very different attitude if you believe it is a fair or an unfair way to evaluate your ability. And your frame of mind can affect how well you perform in the test. So let's see how fair reasoning tests really are.

Who designs reasoning tests?

Reasoning (or psychometric) tests are designed by psychometricians. And if you are wondering if these people need to be the Einsteins of our times, the answer is: not necessarily. Of course it is easier for people to create reasoning tests if they have a high reasoning ability, but it is also very important for psychometricians to be systematic, to generate ideas and to have patience. That's because they need to make sure that what they have created will work.

Reliability and validity

Reasoning tests, like everything else in this world, can be good or bad. In theory anyone could create a test and claim that it measures reasoning ability. It is comforting to know, however, that organisations will not use just anyone's test for their selection system – they will require some proof that the test does indeed measure what it claims to measure. This proof is what we call the test's reliability and validity.

Are reasoning tests reliable?

A test is said to be reliable when it measures something consistently. This means two things. The first is that all the questions within a test should measure some aspect of the same ability. If, for example, the test is a numerical reasoning test, it must measure different aspects of numerical skills.

The second meaning of reliability is that the measurement should be consistent over time, so if the same person were to sit the same test on two different occasions, the results would be more or less the same. There will be some deviation in the results, because there are factors that might affect the score, such as tiredness and concentration. But these external factors affect individuals and not groups of people. So a test is deemed reliable if it consistently measures performance in a large enough sample of people. How large the sample size must be, and how consistent the measurement, is too much information for the purpose of this book, but if you want to learn more about it, a good book to read is *Reliability and Validity Assessment (Quantitative Applications in the Social Sciences)* by Edward G. Carmines and Richard A. Zeller.

Are reasoning tests valid?

The second psychometric property that a reasoning test must have is validity. A test is said to be valid if it actually measures what it was intended to measure. So, a logical reasoning test is valid if it measures a person's logical reasoning ability. This means that if a person takes two different logical reasoning tests, they should get equivalent scores. The drawback is that in order to judge the validity of a test, the test it is compared to must also be valid.

The other way of measuring whether or not a test is valid is by seeing if it predicts the outcomes that it is supposed to. We know, for example, that people who receive higher scores on logical reasoning tests tend to perform better at specific jobs. So if you want to find out if a logical reasoning test is valid, you can see if people who score highly on that test also perform better at those jobs.

This is a very concise description of test reliability and validity. There are in fact several aspects of these properties, and tests have to meet all of them in order to be considered fair. Although there are no legal requirements for a test's reliability or validity level, organisations are not likely to use invalid or unreliable tests as this would defeat the purpose of using them. Moreover, organisations pay a lot of money to use reasoning tests so that they can identify people who will perform well at particular jobs. They choose to use tests that work – tests that are fair.

Discrimination

Apart from the psychometric properties of a test, there is another aspect of whether the test is a fair one. This has to do with whether the test discriminates against people. We must, however, put things into perspective here. Of course reasoning tests discriminate – this is what they are designed to do. They have to discriminate between people who have high or low reasoning ability, and who will therefore be better or worse at the job. So the real question is: do reasoning tests discriminate unfairly?

We have said that it is in the organisation's best interest to use tests that are valid and reliable. But society has made it the state's responsibility (in some countries) to ensure that selection systems, such as psychometric tests, do not unfairly discriminate against minority groups. In the UK for example, there are a number of Acts and Regulations that make it illegal for employers to unfairly discriminate against groups of people in their selection systems, such as the Race Relations Act (1976), the Sex Discrimination Act (1986), the Disability Discrimination Act (1995), the Employment Equality Regulations (2003) and the Age Discrimination Act (2006). These Acts and Regulations protect candidates from being discriminated against due to their ethnic background, gender, disability and age, amongst other things. Going against one of these Acts, for example by rejecting a candidate for a role on the basis of their gender, would be considered direct discrimination. This is illegal in countries such as the UK.

Indirect discrimination can also occur in selection systems, and this can be considered illegal if it is unjustified. Indirect discrimination is when an organisation deselects individuals based on a characteristic which is not directly discriminatory (such as ethnicity or gender), but which indirectly causes discrimination against a protected minority group. For example, if an organisation selects individuals according to their height, they may indirectly discriminate against women, who are typically shorter than men. This would be illegal. If, however, height were used as a criterion to select basketball players, this would be justified (because it affects performance) and would therefore be legally acceptable.

Summary

In this introduction to reasoning tests, we have gone through what these tests are, what types of tests exist, where they originated from, and why and whether you should trust them. We have also described how reasoning tests are used by organisations, and how many people are typically deselected through them. You have now met your opponent.

How will this help you? You should never 'go into battle' without knowing what you are going to face. If you do, you will be anxious and suspicious and will have a negative attitude. None of these will help you win. Even though none of the questions in the test will ask you what reliability or validity means, or whether tests are discriminatory, you will be in a better frame of mind knowing that the test must be reliable and valid, and doesn't discriminate unfairly.

The next step is to get ready for your opponent.

Step 2: Get ready for your opponent

Testing conditions

In the first chapter you were introduced to your opponent. In the chapters that follow, you will have the opportunity to learn the rules behind questions in numerical and logical tests so you will be ready to master your opponent. But apart from practising and improving your numerical and logical reasoning skills, you should also get ready for your opponent in a more practical way. Let's take a step back and go through how you can get ready physically to face your opponent.

So, you have been invited to sit a test for the job position you are interested in. Congratulations! What happens now? There are two possibilities: you may need to take the test at the company's site, or you may be asked to complete the test online.

Taking a test in their field

If you are asked to take the test 'in their field', you will be given the date, time and place of when and where you are expected to sit the test. Make sure that you've found out exactly where it is and how to get there by the day before the test, so that you don't get anxious looking for it and risk a late arrival. Be there half an hour before the test starts. That is enough time to familiarise yourself with the environment and the other people sitting the test, but not enough time to leave you wondering how well you will do, if you are more capable than the rest, or what will happen if you freak out.

Make sure you have all necessary equipment with you such as proof of identification, pen, pencil, eraser, watch and calculator (if permitted). Don't forget things that you may need, such as reading glasses. If you are allowed, take some water with you, though don't get distracted or waste time drinking unless you need to. Remember that time is really important in these tests so you must make the most of it.

Once you are seated, an administrator will make sure you are the person who you say you are by checking your identification. The administrator will then give you instructions on the testing procedure. They will usually explain what type of test you are about to take, how to complete it, how much time you have, what to do when you finish and what the next steps will be. Make sure you listen very carefully and fully understand what is said. If there is anything that isn't clear, do not hesitate to ask, no matter how silly it may seem at the time. But always ask for clarification before the test starts. Even if your mind drifted when the administrator was speaking and you are afraid that you will ask them to repeat something they have already said, it is better to risk annoying the administrator than to be unclear about something regarding the test. Once the test starts, the administrator might not be allowed to answer any questions.

You need to ensure that you are responding precisely according to the test instructions. If you complete the test on paper, it is quite likely that your test will be scored by a computer. This means that if you are asked to put a cross over the correct response, this is exactly what you must do, otherwise you may lose points even though you knew the answer.

During the test, try to stay focused on your paper. Do not worry about how fast other people are progressing and do not pay attention to the administrator, who might be pacing up and down.

Bear in mind that you should never give more than one answer per question, unless you are asked to do so. This will typically be pointed out in the instructions and they are not just doing it to waste ink. If your test is scored by a computer, this will be automatically scored as a mistake. Even if it is scored by a person, they will count it as a mistake, even if one of the two responses is correct (and they are likely to be annoyed by your trying to be cheeky!).

Keep track of time so that you don't find yourself only halfway through when the administrator informs you that you have five minutes, or even just one minute, left. Administrators are very strict about making sure that everyone stops writing when they announce that the time is up. This is to ensure that no participant has an unfair advantage.

Taking a test in your field – online tests

If you have been asked to take a test online, the good thing is that you get to choose the time of day that best suits you and (usually) the day on which you'd like to take it, and you can also create the ideal conditions for you personally. The drawback is that there is no administrator to answer your questions.

Choose wisely

If you are not given a specific day on which you must complete the test, choose the one that suits you best. Perhaps one when there are no other people in the house who might distract you, or the Sunday when the construction workers from the building site across the road have the day off. Avoid leaving this until the last day of your deadline, as something could come up that you can't predict, such as illness or something that could affect your mood. It is always good to allow yourself the flexibility to postpone taking the test if necessary.

Choosing the time should not be difficult. You should choose the time of day when you are most alert, and not prone to tiredness or sleepiness. For example, do not choose to complete the test right after lunch because most people feel a bit sleepy at that time, and don't start the test late at night because you will probably be tired.

Create the ideal conditions

Create conditions which you find comfortable in the room where you will be completing the test. Make sure the temperature is not too hot or too cold. If it is too hot you may be less alert, and you definitely don't want that. It is better to set the temperature a bit cooler, but not so cold that you will feel uncomfortable, as this will disrupt your attention and might force you to interrupt the test to put the temperature right. Remember there is no 'pause' during the test. The clock is ticking and you might be losing valuable time. The room should also be sufficiently lit, but bear in mind that if there is too much light, your eyes might start to hurt after a while.

Consider everything that might distract you. For example, do not have the radio or the TV on. Ask the people in the house not to disturb you for any reason during the test. Take the phone off the hook, and switch off your mobile. Do not start the test on an empty stomach: the feeling of hunger, combined with the noise from your insides, can be very effective distractions. Wear comfortable clothes – tight trousers or an itchy woollen polo-neck top will definitely take your mind off the test.

Smoking is also out of the question during these tests, even if you truly believe that it will help ease your anxiety. The actions of lighting a cigarette, tapping it into the ashtray, inhaling the smoke and putting it down will shift your attention away from the test.

What if you have questions?

Remember that not having an administrator with you in the room has two drawbacks. Firstly, you do not have someone to warn you when time is running out. There might, however, be a countdown watch on your screen which will be very helpful. Secondly, and most importantly, there is no one there to answer any possible questions you might have.

But don't get stressed about this. The people who prepared the online test will have taken this into account and will have made sure that all possible questions are answered in the instructions. It is very important that you read the instructions carefully, taking as much time as you need to fully comprehend them. You can even use a dictionary if there are words you don't understand. If you still have a question, you should contact the company that has asked you to sit the test. Be careful, though – it doesn't look too good if you can't understand detailed instructions that have been put together with particular care so that participants won't have to call the company.

Practice questions

Most tests will typically have some practice questions before the start. Take as much time as you need to get them right. If you made mistakes, find out why. If you are taking the test at the recruiter's site, ask the administrator to explain anything that is not clear. If you are at home, take as much time you need to understand the mistake. If you feel that you could practise some more and you have the time to do so, defer the test for another day.

Check your computer specifications

Finally, you have to make sure that the specifications suggested by the company match your computer. If not, you will probably be unable to sit the test and will have to find another computer. The same applies to the internet: if you have a very slow internet connection, you might not be able to complete the test.

The companies which create the tests try to ensure that the vast majority of candidates will not have any difficulty completing the test online. If there are computer or internet specifications, these will be detailed in the invitation email that you have received, or they will be mentioned in the instructions when you enter the testing site. You can also 'test' whether you have any accessibility issues when you complete the practice questions of the test.

A problem may arise if your pop-up blockers are active, and the test you have to take uses pop-ups. In this case you will probably receive instructions on how to de-activate your pop-up blockers. If you don't receive any instructions, and you can't solve the problem yourself, you can contact the testing company for help. There should always be a contact number or email address for technical problems.

Do not, however, contact the company and expect them to find a solution for you, unless you have exhausted every possibility of solving the problem yourself. You don't want to give the impression that you are unable to solve problems that are likely to arise in your line of work, as this might look bad when the company decides whether or not you are the most suitable person for the job. If you do find that you have an issue that you can't resolve, such as your computer or internet connection will not allow you to take the test or will slow you down, you will have to find another computer to take the test on.

Be prepared

What to do when you find out you need to take a test

One of the most important things to do when you find out that you need to sit a test is to begin preparing for it as early as you can. Preparation includes studying practice questions as well as preparing yourself psychologically. The latter involves positive psychology and relaxation techniques, both of which will be discussed later in this chapter (see pages 27–9).

The first thing you need to do is find out as much as you can about the test itself. What type of questions or items are likely to appear?

What is the format of the test? In what form are the questions usually presented? How many multiple-choice answers will you have? How much time will you have at your disposal? You need to get acquainted with the logic behind each psychometric test and then complete all the practice questions. The more familiar you are with the test, the less anxious you will be.

Don't wait until the last moment to study for the test. It might be helpful if you set a schedule as soon as you are asked to sit a test, indicating how many days you would like to spend on preparation and what you will study each day. This is a good tip, especially if you are normally not very organised. You need to find a quiet place to study without too much noise, which can include traffic outside the window, television or radio, and children constantly seeking your attention (the latter may be hard to avoid!). These conditions might not be exactly the same as the ones under which you'll complete the test, but they are the best possible conditions for understanding and absorbing all the information necessary to get to know the test you are about to take.

Practice

Now that you have found out what the test is all about, you have organised your schedule and you have found the ideal place to study, there's only one thing left to do. Practise! As mentioned numerous times already, practice can and will make you better at psychometric tests. The more you practise, the higher your final score will be.

In the first chapter we explained why it is most likely that you'll be asked to take a speeded (or timed) test. This means that you will have two conflicting demands: you need to work quickly as well as accurately. There is, of course, the paradox that the less time you spend checking details to ensure accuracy, the less time you have for solving additional questions. The line between speed and accuracy is quite thin, and you will need to find the balance that works best for you. This will come intuitively the more you practise solving reasoning tests.

You must also learn to pace yourself when taking numerical and logical practice tests. When half the time is up, check whether you are halfway through the test. If you aren't, try to work faster or to spend less time double-checking your answers. When you take the real test, make sure you know how much time you have to complete the test, and try to pace yourself as you go through it.

The final stage of preparation is to take a practice test under real conditions. Take one at the same time of day that the real test will take place, so that you can see how tired you are and what noise or other distractions might influence your performance.

Small things that make a difference

There are other small things that you can do a few days before the test, and most definitely on the day of the test, to keep you feeling good physically. This will help you to achieve your maximum performance.

Sleep

The most important thing to do is to get enough sleep the week before the test, especially the night before. Reducing your sleep to spend the extra hours preparing for your test will have a negative effect on your performance. The ideal amount of sleep for adults is said to be eight hours. If you have eight hours' sleep, you won't have a sleep deficit, so you'll be in a good mood and in a better position to concentrate.

Food

You should also eat well, and try especially hard to have a good and nutritious breakfast on the morning of the test. This can have an excellent effect on your body and mind. Also, having a power drink that is high in vitamins, or some chocolate to give you a glucose boost just before you take the test, can help prepare your body to work at maximum efficiency. Bear in mind that some people experience a drop in energy about half an hour after eating a sugar snack. If you are one of these people, check how long the test will take and calculate when would be the best time to boost your sugar levels to avoid having a 'sugar slump' while taking the test.

Exercise

You can help your body to be more efficient during the test by doing some exercise. Exercising on the day of the test might tire you out, but try to incorporate some into your routine for the week leading up to the test. Remember that your brain is part of your body, and it will function better if your body is in a better state.

Illness

This also means that if you are ill, and especially if you have a fever, you should try to postpone taking the test. If you are invited to take the test on a specific date this may not be easy to change. You should make a judgement call there – if you feel that your illness affects your concentration, your thinking processes and your reflexes, it would be better to postpone the test, rather than perform less efficiently. If, on the other hand, you just have a runny nose, it would be better not to ask the company to reschedule your test at the last minute.

Chemicals

You must bear in mind that you should limit any chemical usage, certainly drugs but also alcohol, when you are about to take a reasoning test. This is because chemicals affect your thinking processes and therefore your performance when taking a numerical or a logical test.

Take the winning approach

We've discussed how you can prepare physically to perform to the best of your ability when taking a psychometric test. It is equally important, however, to prepare yourself mentally. You need to put yourself in the right frame of mind to give your best performance. This brings us to the concept of positive psychology – taking the winning approach.

Positive psychology

Positive psychology is a relatively recent branch of psychology that focuses on characteristics and strengths which give us the potential to thrive. A key factor of positive psychology is confidence – the trust we have in our abilities. Confidence is so relevant to what we achieve in life, that it is a key term used by people involved in sport, enterprise, business and public speaking. The same principle of having a positive attitude and being confident in your abilities can have a very positive effect when you take a psychometric test.

Of course confidence in your abilities will gradually build as you practise solving psychometric tests, but you must also work at it. Mental preparation is one way of improving your positive psychology: visualise sitting the test and performing really well. This will help you perform better, because anxiety is created by what you expect, or what you think will happen. If you think you will perform poorly, this produces a corresponding negative emotional reaction (essentially stress or anxiety) and as a result you are more likely actually to perform poorly.

If you're feeling anxious a few days before the test, take a moment to consider what it is exactly that you're feeling anxious about. Is it because you have not studied enough, or in sufficient depth, or because you are not feeling sure of what the test is about? If this is the case, then you haven't done enough preparation, or the right kind of preparation, for the test. Luckily, you have a few more days. Plan a schedule carefully and follow it step by step.

Or are you feeling anxious because you have failed previous tests, because you fear the other candidates will perform better or because you suddenly suffer from 'blank outs'? If this is the case, you need to work on your relaxation techniques.

Relaxation techniques

Relaxation is one of the most important things to practise in order to do the test well, because stress can impair your memory. The more agitated and anxious you are, the more it will affect your performance and the worse you will do. Relaxation exercises are easy to learn and are very effective in dealing with stress and test anxiety.

There are a number of things you can do in order to calm your nerves. Breathing is one of the easiest methods. Breathe deeply, expanding your belly on the inhale, and then let the stress out by exhaling slowly. Do this as many times as you need to in order to feel relaxed.

Another way of relaxing is by thinking about, and then relaxing, every muscle of your body. You need to concentrate on each muscle separately, starting from your neck muscles, moving down to your arms, then your back, and then moving progressively down your body all the way to your toe muscles. A quicker relaxation exercise is to make a fist and then release it slowly.

Such relaxation exercises require practice twice a day for two to three weeks until you become an 'expert' in quickly and effectively reducing your anxiety level. It is not wise to practise them in a room with distractions, such as noise or other people, or when you're feeling hungry or sick, or when you've been exercising and so are feeling full of energy. The best position to practise the techniques is sitting comfortably, with your legs crossed and your arms resting by your sides.

Once you finish your relaxation exercise, do not immediately jump up and run around the room filled with happiness that you managed to relax! Sit still for a couple of minutes and give your body time to return to its normal state.

Once you have picked your preferred relaxation technique, practise it for a couple of weeks before the test. When you are ready to take the test, take a few minutes to go through the technique and relax. Now you are ready to face your opponent.

Step 3: Master your opponent – numerical tests

People say that mathematical ability is one of those things: you either have it or you don't. Although it is true that some people find it easier than others to deal with maths, your ability can also be a 'self-fulfilling prophecy'. So if, for example, your maths teacher made a premature judgement that you were not good at maths, this could have got you into a vicious circle of feeling less confident about your ability, causing you to spend less time working at maths, therefore building weaker foundations and consequently performing poorly in class and essentially reinforcing your teacher's assumption.

However, the important thing is whether or not you created good foundations. Maths is not like history, where you could skip a topic and still learn everything else with the same ease. In maths, every topic is built on previous blocks. You can't build a strong house on weak foundations and the same goes for building your knowledge in maths and, consequently, in numerical reasoning tests.

We will therefore start by taking you back to the basics. If you find you already have strong foundations you can move on to the next section, but have a flick through these pages first, just in case.

Basic calculations

As you may remember, the basic calculations are addition, subtraction, multiplication and division. You will typically be allowed to use a calculator in a numerical test, which means you probably won't be asked to do these basic calculations in your head.

However, most numerical tests are timed. You will therefore need to answer quickly as well as accurately. It would be useful to refresh the skill of solving basic calculations, as it will save you time if you can do these in your head instead of using a calculator. Here are a few examples:

Addition

	QUESTION	ANSWER
1.	34 + 27 =	
2.	2.18 + 21.03 =	
3.	78 + 24 =	
4.	299 + 321 =	
5.	0.88 + 0.22 =	

Subtraction

	QUESTION	ANSWER
6.	127 − 45 =	
7.	88 − 29 =	
8.	1 − 0.89 =	

9. $444 - 27 =$

10. $32 - 56 =$

Multiplication

QUESTION	ANSWER

11. $7 \times 8 =$

12. $6 \times 9 =$

13. $5 \times 12 =$

14. $4 \times 14 =$

15. $3 \times 13 =$

Division

QUESTION	ANSWER

16. $32 \div 4 =$

17. $48 \div 12 =$

18. $72 \div 8 =$

19. $49 \div 7 =$

20. $36 \div 2 =$

Rounding off numbers

Many of the calculations you will make will not give you a rounded answer. Multiple-choice options in numerical tests, however, are usually rounded off. You must therefore be familiar with how to round off numbers so that you can choose the correct answer.

The rule behind rounding off is that digits from 0–4 are rounded down and digits from 5–9 are rounded up. For example, 1.0, 1.1, 1.2, 1.3 and 1.4 are all rounded off to 1. On the other hand, 1.5, 1.6, 1.7, 1.8 and 1.9 are all rounded off to 2.

What you also need to consider is how many digits you have, and what digit you need to round off to. When you are rounding off, you only take into account the first digit on the right.

Take, for example, the number 1.354. If you need to round this off to an integer (i.e. a whole number), you only consider the digit of the first decimal place (i.e. on the right of the decimal point), which is 3 in this example. So 1.354 would be rounded down to 1.

If you wanted to round off the same number to one decimal place, you would only consider the digit on the right of the first decimal place, which is 5 in this example. So 1.354 would be rounded up to 1.4.

If you wanted to round off the same number to two decimal places, you would only consider the digit on the right of the second decimal place, which is 4 in this example. So 1.354 would be rounded down to 1.35. And so on.

Practice questions
Round off the following numbers to one decimal place.

	QUESTION	ANSWER
21.	13.472	
22.	1.54	

23. 1.45

24. 2.3499

25. 0.991

Round off the following numbers to two decimal places.

	QUESTION	ANSWER
26.	24.843	
27.	145.4451	
28.	4.2982	
29.	0.8151	
30.	2.0016	

Transforming numbers and percentages

Percentages are used to express how large one quantity is relative to another quantity. In maths, a percentage is a way of expressing a number as a fraction of 100 ('per cent' meaning 'per hundred'). It is often denoted using the sign '%'. For example, 24% (read as 'twenty-four per cent') is equal to $^{24}/_{100}$, or 0.24.

In the following sections you will often need to turn numbers quickly into percentages and vice versa in order to solve a question. It will save you a lot of time if you can do these transformations in your head rather than using a calculator.

Practice questions

Transform the following numbers into percentages.

	QUESTION	ANSWER
31.	0.17	
32.	0.242	
33.	4.1	
34.	1.02	
35.	100	
36.	0.023	
37.	1.98	

Transform the following percentages into numbers (give your answer to three decimal places).

	QUESTION	ANSWER
38.	29%	
39.	35.5%	
40.	42.1%	
41.	0.8%	

42. 0.27%

43. 100%

44. 120%

45. 1000%

46. 106.84%

Write the following fractions as percentages, without using a calculator.

	QUESTION	ANSWER
47.	½	
48.	⅕	
49.	²⁄₈	
50.	¹²⁄₆	
51.	⁴⁰⁄₅₀	
52.	²·⁵⁄₅	
53.	⁴⁵⁄₉	
54.	¾	
55.	⁵⁰⁄₂₀₀	

Learn the rules of numerical reasoning

Most of the questions in numerical reasoning tests are based on simple calculations, which usually involve averages, percentages, ratios and unit conversions. In this section we will take you through the kinds of questions that are typical in numerical reasoning tests, and some rules that you can follow to solve these. We will also look at some 'tricks' that psychometricians use to make these questions more complex.

Averages

Although there are four types of average (the mean, mode, median and range), questions in numerical reasoning tests typically refer to the mean, and the term 'mean' is used interchangeably with the term 'average'. The mean of several numbers is calculated by adding up all their values and dividing the result by how many numbers there are.

Questions on averages usually appear in one of two ways. Here is the first:

GENERAL QUESTION: What is the average of X, Y and Z?

GENERAL SOLUTION: Add the values together and divide the total by the number of values.

FORMULA: $(X + Y + Z) \div N$ (where N is the number of values)

EXAMPLE QUESTION: What is the average of 13, 18, 10 and 11?

EXAMPLE SOLUTION: $(13 + 18 + 10 + 11) \div 4 = 52 \div 4 = 13$.

Practice questions

....

QUESTION   **ANSWER**

56. What is the average of 257, 215 and 197?

     a.   216
     b.   219
     c.   221
     d.   223
     e.   227

....

QUESTION   **ANSWER**

57. What is the average of 1.6, 1.07, 2.6, 4.22 and 3?

     a.   2.1
     b.   2.5
     c.   2.7
     d.   2.9
     e.   3.1

....

QUESTION   **ANSWER**

58. What is the average of 27, 44, 39, 25 and 43?

     a.   32.8
     b.   33.1
     c.   35.6
     d.   37.2
     e.   38.7

	QUESTION	ANSWER

59.   What is the average of ¼ , ³⁄₆ and ⅘ ?

    a.   0.30

    b.   0.35

    c.   0.38

    d.   0.40

    e.   0.52

These calculations are quite straightforward. But they are unlikely to be included in a numerical test in such a straightforward way. There are two ways in which these calculations can be made more 'complex'. The first is by giving you actual examples rather than just numbers.

Practice questions

	QUESTION	ANSWER

60.   Three boxes have a capacity of 35 cm², 45 cm² and 1x10 m². What is the average capacity of a box?

    a.   26.03 cm²

    b.   60.3 cm²

    c.   61.7 cm²

    d.   63.3 cm²

    e.   63.6 cm²

QUESTION	ANSWER

61. A tour guide had five tours on Monday. The groups he had were 20, 15, 8, 13 and 9 people. What was the average size of his tour groups on Monday?

    a.   11

    b.   12

    c.   13

    d.   14

    e.   15

QUESTION	ANSWER

62. A student received the following grades in his exams: Maths 67%, Chemistry 50%, Physics 74%. What was his average grade in his exams?

    a.   63%

    b.   64%

    c.   65%

    d.   66%

    e.   67%

The second way in which questions can be made more complex is by presenting the information in a table, graph or chart. This requires an additional mental step of finding the information you need before you make the calculation. The following questions are examples of this.

FIGURE 1: **Berlin weather information**

		JAN	FEB	MAR	APR	MAY
Temperature (°C)	MINIMUM	−1.9	−1.5	1.3	4.2	9
	MAXIMUM	2.9	4.2	8.5	13.2	18.9
Mean total rainfall (mm)		42.3	33.3	40.5	37.1	53.8
Mean number of rain days		10	8	9.1	7.8	8.9

Practice questions

....

QUESTION   ANSWER

63. What was on average the maximum temperature
in Berlin between January and May?

    a.   9.5°C

    b.   10.1°C

    c.   10.4°C

    d.   10.9°C

    e.   11.5°C

....

QUESTION   ANSWER

64. What was the daily average temperature in
Berlin in February?

    a.   1.2°C

    b.   1.4°C

    c.   1.6°C

d.   1.8°C

e.   2.9°C

---

65.   On average, how many days of rainfall were
there in Berlin between February and April?

    a.   8.1

    b.   8.3

    c.   8.4

    d.   8.5

    e.   8.6

---

66.   What was the mean total rainfall between
January and May?

    a.   39.4mm

    b.   39.7mm

    c.   41.2mm

    d.   41.4mm

    e.   42.3mm

As we have mentioned, questions on averages can also involve a different
scenario.

GENERAL QUESTION: If the average of X, Y and Z is A, what is the value
of X, if you know the values of Y, Z and A?

GENERAL SOLUTION: Multiply the average (A) by the number of values
that make up the average (i.e. here the number of values was 3,
since you have X, Y and Z). From this 'total', you then subtract

the values that are given (i.e. Y and Z). The remaining value is the value of X.

FORMULA: $(A \times \text{number of values}) - Y - Z$

EXAMPLE QUESTION: If the average of 4, 16, 22, 27 and X is 15, what is the value of X?

EXAMPLE SOLUTION: You need to multiply 15 by 5 (the number of values) and then subtract the values that are given: $(15 \times 5) - 4 - 16 - 22 - 27 = 75 - 4 - 16 - 22 - 27 = 6$.

Practice questions

QUESTION	ANSWER

67. What is X, if the average of 230, 255, 210 and X is 240?

   a.   257

   b.   259

   c.   263

   d.   265

   e.   268

QUESTION	ANSWER

68. What is X, if the average of 1.5, 1.2, 1.9 and X is 2?

   a.   2.5

   b.   2.7

   c.   2.9

d. 3.1

e. 3.4

QUESTION   ANSWER

69. What is X, if the average of 13, 19, 21, 14 and X is 16?

    a. 13

    b. 14

    c. 15

    d. 16

    e. 17

FIGURE 2: **Revenues per quarter**

	Company A	Company B	Company C
Quarter 1	£3,000,000	£4,200,000	£4,300,000
Quarter 2	£3,500,000	£4,400,000	£5,600,000
Quarter 3	£3,300,000	£5,400,000	?
Quarter 4	£3,600,000	?	£6,700,000

QUESTION   ANSWER

70. What was on average the quarterly revenue of company A?

    a. £3,300,000

    b. £3,350,000

    c. £3,360,000

    d. £3,380,000

    e. £3,400,000

QUESTION   ANSWER

71.   What was on average the Quarter 2 revenue of
      the three companies presented?

      a.   £4,300,000
      b.   £4,400,000
      c.   £4,500,000
      d.   £4,600,000
      e.   £4,700,000

QUESTION   ANSWER

72.   What was the Quarter 3 revenue of Company C,
      if on average the Quarter 3 revenue was
      £4,900,000?

      a.   £5,200,000
      b.   £5,450,000
      c.   £5,900,000
      d.   £6,000,000
      e.   £6,200,000

QUESTION   ANSWER

73.   What was the Quarter 4 revenue of Company B,
      if on average the quarterly revenue of Company
      B was £4,875,000?

      a.   £5,100,000
      b.   £5,200,000
      c.   £5,300,000
      d.   £5,400,000
      e.   £5,500,000

There are essentially two things you need to practise in order to improve your ability to solve numerical reasoning questions. The first is to practise *making the required computations quickly and accurately*. The more times you solve, for example, questions on averages, the more the generalised solution will become hard-wired in your head. So that when you are asked to compute the average of a set of values, you can immediately add the values and divide by the number of values, instead of spending time thinking about – or trying to remember – the solution.

The second thing you need to practise is *identifying what kind of computation each question requires*, so that you can immediately use the correct one instead of going through a trial-and-error process. As we have mentioned, questions are made more complex by masking them with wording, or by including the data in tables and charts. The more questions you solve, the easier it will become to identify what type of calculation you need to do in order to solve a question.

This book will help you with both aspects of practising: with identifying the type of computation required, and with making the computations quickly and accurately. You are therefore given some questions where you can focus on learning the calculations behind a generalised solution, and other questions where you need to identify what kind of computation to use.

Percentages

As we have already seen, percentages are used to express how large one quantity is relative to another quantity. Percentages can be shown in a number of ways in a numerical test. We will help you become familiar with the different types of questions, take you through the process of solving them and give you examples to practise.

When performing calculations with percentages you should remember that the percentage can be treated as being equivalent to $\frac{1}{100}$ or 0.01. For example, 15% of 200 can be written as $(15 \div 100) \times 200$, and also as 0.15×200. Both these ways of presenting the percentage give you the same answer: 30.

GENERAL QUESTION: What is A as a percentage of C?

GENERAL SOLUTION: Divide A by C. To get a percentage, multiply this number by 100 per cent.

FORMULA: $(A \div C) \times 100\%$

EXAMPLE QUESTION: What percentage of a class are girls, if there are 35 students in the class and 13 of these are girls?

EXAMPLE SOLUTION: You need to divide the number of girls by the total number of students. This will show you how big one group is relative to the other. The answer is $13 \div 35 = 0.37$.

Of course when you divide two numbers, the solution will be a number and not a percentage. All that remains to do is to transform that number into a percentage (by multiplying by 100 and adding the % sign). Here, turning 0.37 into a percentage gives you the answer of 37%. This question can be asked with any type of group or numeric quantity.

One way that psychometricians might try to make a question a bit more complex is by adding an extra step to the calculations you need to make, before you get to the solution.

EXAMPLE QUESTION: What percentage of a class are *boys*, if there are 35 students in the class and 13 of these are girls?

EXAMPLE SOLUTION: You need to calculate the number of boys in the class, and then find what percentage that is of the total number of pupils.

You can find the number of boys by subtracting the number of girls from the total number of pupils, i.e. 35 − 13 = 22. You can then calculate the percentage by dividing the two quantities (22 ÷ 35 = 0.629) and then transforming the result into a percentage by multiplying it by 100 and adding the % sign (i.e. 0.629 × 100% = 63%).

You could also get to the same solution by calculating the percentage of girls in the class (i.e. 37%) and subtracting this from 100%. This is because boys and girls make up 100% of the class, so if you take out the percentage of girls you will be left with the percentage of boys. Both calculations take the same number of steps, so just use the first of the two that comes to mind.

Practice questions

....

QUESTION   ANSWER

74.   What is 35 as a percentage of 200?

    a.   12.5%

    b.   15%

    c.   17.5%

    d.   20%

    e.   22.5%

....

QUESTION   ANSWER

75.   What is 0.25 as a percentage of 10?

    a.   0.025%

    b.   0.25%

    c.   0.5%

    d.   2.5%

    e.   5%

QUESTION	ANSWER

76. What is 12 as a percentage of 6?

    a.   50%

    b.   125%

    c.   150%

    d.   200%

    e.   250%

QUESTION	ANSWER

77. During the previous year (which was not a leap year) it rained on 61 days. What percentage of that year did it rain?

    a.   9%

    b.   11%

    c.   13%

    d.   17%

    e.   21%

QUESTION	ANSWER

78. The total area of the UK is 244,820 km^2, of which 3,230 km^2 is water. What percentage is the land area of the UK?

    a.   96.4%

    b.   96.9%

    c.   97.2%

    d.   97.6%

    e.   98.7%

79.   If Canada has 1,042,300 km of roadways, of
          which 415,600 km are paved, what percentage
          are the unpaved roadways in Canada?

    a.   39.9%

    b.   57.7%

    c.   59.8%

    d.   60.1%

    e.   61.5%

These examples asked you to turn numbers into percentages. Now let's
look at another type of question on percentages: turning percentages
into numbers.

GENERAL QUESTION: What is x% of A?

GENERAL SOLUTION: A multiplied by x%.

FORMULA: A × (x ÷ 100)

EXAMPLE QUESTION: If a company has 460 employees, of which
5% are working part time, how many part-time employees does
the company have?

EXAMPLE SOLUTION: In order to find 5% of 460, you need to
multiply 5% by 460. Remember that 5% is just 5 ÷ 100. So the
calculation is: (5 ÷ 100) × 460 = 23.

TIP: If you need to make a calculation (such as a multiplication) with
a percentage, the quickest way to do this is by transforming the
percentage into a number, i.e. transforming 5% into 0.05. The multiplication then becomes 0.05 × 460 = 23. How much time will this

save you? It will save you the seconds that it would take to press the additional buttons on the calculator (to divide 5 by 100). Is it worth it? Yes. You will probably need to make a number of these calculations during a numerical test. Any time you save will be time that you can use to solve another question and gain extra marks.

Practice questions

....

QUESTION	ANSWER

80. What is 24% of 36?

    a.   8.64
    b.   9.35
    c.   10.29
    d.   11.05
    e.   12.90

....

QUESTION	ANSWER

81. What is 0.5% of 200?

    a.   0.5
    b.   1
    c.   1.2
    d.   5
    c.   10

....

QUESTION	ANSWER

82. What is 120% of 40?

    a.   42
    b.   44
    c.   46

d.   48

e.   50

....

83.   If Canada has a labour force of 17.95 million,
and 76% of these people work in services, what
is the total population working in services in
Canada?

    a.   12.71 million

    b.   13.64 million

    c.   13.98 million

    d.   14.02 million

    e.   14.56 million

....

84.   If a shirt's original price is £35, and it has a 20%
discount, how much money would you save if
you bought it during the discount period?

    a.   £7

    b.   £8

    c.   £9

    d.   £10

    e.   £11

....

85.   Italy has a population of 58,145,320 individuals.
If 13.6% of these are aged 0–14 years, how many
people are aged over 14 in Italy?

a.   50,237,556
b.   50,693,241
c.   51,023,088
d.   51,109,776
e.   51,231,505

Note that the last example involved the extra step of calculating the number of people aged over 14 years rather than the number of people aged 0–14 years.

There is another extra step that may be added to make a calculation using percentages more difficult: having to add the percentage increase on to the original amount.

GENERAL QUESTION: What is A increased by B%?
GENERAL SOLUTION: Find what B% of A is, and then add this on to A.
FORMULA: $(A \times B\%) + A$

EXAMPLE QUESTION: If you get a 3% increase in your current £27,000 salary, what will your new salary be?

EXAMPLE SOLUTION: You need to find 3% of 27,000 (the increase you will receive), and the extra step is to add this to your current salary, in order to get the total new salary. The answer is: $(0.03 \times 27,000) + 27,000 = 810 + 27,000 = 27,810$. (Here we transformed 3% into 0.03 as a shortcut in making the calculation.)

TIP: There is another shortcut you can use when calculating a percentage increase of a value. Instead of multiplying 0.03 by 27,000 and then adding that to 27,000, you can make just one calculation: $1.03 \times 27,000 = 27,810$.

As you can see, this gives you exactly the same result as $(0.03 \times 27{,}000)$ + 27,000. Why is that? Because this calculation can be transformed as follows:

$$(0.03 \times 27{,}000) + 27{,}000 =$$
$$(0.03 \times 27{,}000) + (1 \times 27{,}000) =$$
$$(0.03 + 1) \times 27{,}000 =$$
$$1.03 \times 27{,}000$$

These steps are only given to show you how the shortcut works. All you really need to know is the shortcut itself: when you have to calculate a percentage increase of a value, turn the percentage into a number and add 1. Then multiply this by the value.

Practice questions

QUESTION	ANSWER

86.   What is 120 increased by 11%?

    a.   131
    b.   131.2
    c.   132
    d.   132.2
    e.   133.2

QUESTION	ANSWER

87.   What is 3,000 increased by 0.1%?

    a.   3,001
    b.   3,003
    c.   3,010

d. 3,030

e. 3,300

QUESTION   ANSWER

88. What is 0.5 increased by 0.5%?

    a.   0.503

    b.   0.515

    c.   0.520

    d.   0.525

    e.   0.750

QUESTION   ANSWER

89. If the average temperature in Athens was 25°C in June, and increased by 12% in July, what was the average temperature in July?

    a.   26°C

    b.   27°C

    c.   28°C

    d.   29°C

    e.   30°C

QUESTION   ANSWER

90. If 300 televisions were sold this month, and the target for next month is to increase television sales by 2%, how many televisions must be sold to reach that target?

    a.   302

    b.   304

    c.   306

d.   308

e.   310

You may be asked to compute an increase that is over 100%. You would still go through the same calculations to find the answer. So if you are asked 'How many people will be employed by a company of 3,000 employees if its workforce increases by 120%?', you would calculate: $3,000 \times (1 + 120\%) = 3,000 \times (1 + 1.2) = 3,000 \times 2.2 = 6,600$.

Practice questions

....

QUESTION   ANSWER

91.   What is 50 increased by 100%?

    a.   80

    b.   90

    c.   100

    d.   110

    e.   150

....

QUESTION   ANSWER

92.   What is 50 increased by 200%?

    a.   80

    b.   90

    c.   100

    d.   110

    e.   150

93. What will the revenue of Gypas Co. be if it
     increases by 170% from the current £5 million?

     a.   £135,000
     b.   £1,350,000
     c.   £13,500,000
     d.   £135,000,000
     e.   £1,350,000,000

This type of question on the percentage increase of a quantity can also
ask about a successive percentage increase. Take a look at the following
question:

94. The population of Malta in 2008 was 403,000.
     What will the population be in 2011, if the
     population's growth rate is 0.5% per annum?

     a.   405,015
     b.   407,040
     c.   409,045
     d.   409,075
     e.   409,120

You can solve this question by working out the population's increase
consecutively for each year:

   In 2009: 403,000 × 1.005 = 405,015
   In 2010: 405,015 × 1.005 = 407,040
   In 2011: 407,040 × 1.005 = 409,075

So the answer is **d**. As you can see, this is not in fact a more difficult question, it just takes more steps to get to the answer.

The previous examples involved calculations with a percentage increase of a value when *given the initial value*. Another type of question relating to a percentage increase of a value involves finding the initial value when *given the end value*.

GENERAL QUESTION: What was B before it increased by x%?
GENERAL SOLUTION: You need to divide B by (1 + x%).
FORMULA: B ÷ (1 + x%), or B ÷ ((100 + x) ÷ 100)

EXAMPLE QUESTION: If the price of a flat has increased by 5% since last year, and it is currently £189,000, what was its value last year?

EXAMPLE SOLUTION: Remember that when you were given the original value (A) and were asked to calculate the current value (B) after an increase (x%), you had to multiply A by (1 + x%) to get to B. Now that you have B and need to get to A, you need to perform the opposite calculation, i.e. divide B by (1 + x%). So you need to work out 189,000 ÷ (1 + 0.05) = 189,000 ÷ 1.05 = 180,000.

Practice questions

QUESTION	ANSWER

95. What was 165, before it increased by 10%?

   a.   150
   b.   151
   c.   154

d.   155

e.   156

---

96.   What was 21, before it increased by 40%?

a.   12

b.   13

c.   14

d.   15

e.   16

---

97.   What was 703.5, before it increased by 0.5%?

a.   670

b.   673.5

c.   700

d.   700.5

e.   701.25

---

98.   If there were 3,520 earthquakes in the USA this
year, which reflects a 10% increase since last
year, how many earthquakes were there last year?

a.   3,100

b.   3,120

c.   3,150

d.   3,180

e.   3,200

QUESTION	ANSWER

99. If the revenues of a company have increased since last year by 7%, to reach 1.177 million euros, what were the revenues last year?

 a. 1 million euros

 b. 1.02 million euros

 c. 1.07 million euros

 d. 1.1 million euros

 e. 1.11 million euros

We have looked at calculations related to percentage increases in value, given the initial or the end values. Now let's look at how percentage decreases in value work.

GENERAL QUESTION: What is A decreased by x%?

GENERAL SOLUTION: Work out what x% of A is, and subtract it from A.

FORMULA: $A \times (1 - x\%)$, or $A \times ((100 - x) \div 100)$

EXAMPLE QUESTION: If the revenues of a company decreased by 3% from Quarter 2, and the revenues of Quarter 2 were £1.6 million, what were the revenues of the company in Quarter 3?

EXAMPLE SOLUTION: You need to calculate 3% of the Q2 revenues (the decrease), and subtract this from the Q2 revenues. The decrease is 3% of £1.6 million, or $0.03 \times 1,600,000 = 48,000$. If you subtract this from the initial 1,600,000, what remains is £1,552,000.

TIP: As with many other calculations, there is a shortcut that you can take when calculating the decrease of a percentage. You can subtract the percentage decrease from 100% (i.e. in the previous example this would be 100% − 3% = 97%), and multiply this by the total (i.e. £1,600,000 × 0.97). This will give you the same answer: £1,552,000. Is this really a shortcut? After all, you make the same number of calculations: one subtraction and one multiplication; you just make them in a different order. The difference is, if you do the subtraction first, this will always be a subtraction from 100. In most cases you will be able to do this in your head much more quickly than using a calculator. So again it will only save you a few seconds, but they are seconds that can count.

Practice questions

QUESTION   ANSWER

100. What is 350 decreased by 20%?

    a.   250
    b.   260
    c.   270
    d.   280
    e.   290

QUESTION   ANSWER

101. What is 6,000 decreased by 69%?

    a.   1,860
    b.   1,980
    c.   2,000
    d.   2,040
    e.   4,140

| QUESTION | ANSWER |

102. What is 10 decreased by 9%?

   a.   8.9

   b.   9.0

   c.   9.1

   d.   9.2

   e.   9.3

| QUESTION | ANSWER |

103. If a coat costs £120, and there is a 30% sale on all items in the store, how much would you pay for the coat?

   a.   £80

   b.   £84

   c.   £88

   d.   £90

   e.   £100

| QUESTION | ANSWER |

104. If 2,180 washing machines were sold last quarter, and a 5% decrease in sales is expected for next quarter, how many washing machines are expected to be sold next quarter?

   a.   2,071

   b.   2,195

   c.   2,200

   d.   2,270

   e.   2,300

The previous examples involved calculations with a percentage decrease of a value when *given the initial value*. Let's also look at how to find the initial value when *given the end value*. Take the following question:

105. If the revenues of a company decreased by 5% from Quarter 1, to reach £4,987,500 in Quarter 2, what were the revenues of the company in Quarter 1?

    a.   4,738,125
    b.   4,925,600
    c.   5,120,500
    d.   5,250,000
    e.   5,302,600

Remember that when you were given the original value (A) and were asked to calculate the current value (B) after a decrease (x%), you had to multiply A by (1 − x%) to get to B. Now that you have B and need to get to A, you need to perform the opposite calculation, i.e. divide B by (1 − x%). So you can work out 4,987,500 ÷ (1 − 5%) = 4,987,500 ÷ 0.95 = 5,250,000.

Practice questions

....

QUESTION   **ANSWER**

106. If A decreased by 15% to become 120, what was A?

    a.   126.3
    b.   134.2
    c.   135.5
    d.   138.0
    e.   141.2

QUESTION   ANSWER

107. If A decreased by 40% to become 0.48, what
was A?

    a.   0.56

    b.   0.64

    c.   0.72

    d.   0.80

    e.   0.88

QUESTION   ANSWER

108. If A decreased by 2% to become 10,000, what
was A?

    a.   10,020

    b.   10,200

    c.   10,204

    d.   10,402

    e.   12,000

QUESTION   ANSWER

109. If the marketing budget of a company is
£199,500, which is a 5% decrease since last year,
how much was the marketing budget last year?

    a.   £209,000

    b.   £209,250

    c.   £209,500

    d.   £209,750

    e.   £210,000

QUESTION   ANSWER

110. The labour population of a country has
     decreased since last year by 1%, to 2,178,000
     million. What was the labour population last
     year?

   a.   2,190,000
   b.   2,200,000
   c.   2,202,000
   d.   2,215,000
   e.   2,220,000

Now that you have gone through the most common types of questions
on percentages, here are a few additional examples. These are a mixture
of all the different types, so that you can bring together everything you
have learned.

Practice questions

FIGURE 3: **Weather information**

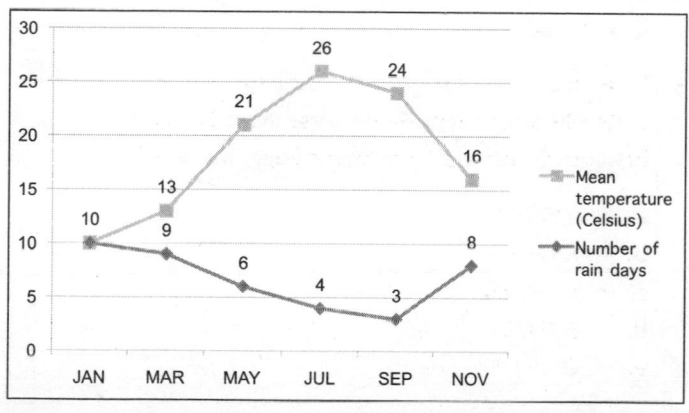

QUESTION	ANSWER

111. What was the percentage increase in the mean temperature from March to May?

    a.   60%
    b.   62%
    c.   64%
    d.   66%
    e.   68%

QUESTION	ANSWER

112. If the temperature from last December to January increased by 11%, what was the temperature last December?

    a.   8°C
    b.   8.5°C
    c.   8.7°C
    d.   9°C
    e.   9.2°C

QUESTION	ANSWER

113. What was the percentage decrease in the temperature from July to November?

    a.   38%
    b.   41%
    c.   44%
    d.   47%
    e.   62%

QUESTION   ANSWER

114. If the mean temperature increases by 5% from
     this March to next March, what will it become?

   a.   13.3°C
   b.   13.5°C
   c.   13.7°C
   d.   13.9°C
   e.   Cannot say

QUESTION   ANSWER

115. If the mean temperature decreases by 15%
     from November to the following January, what
     will the mean temperature be in January?

   a.   11.2°C
   b.   11.9°C
   c.   12.4°C
   d.   12.9°C
   e.   13.6°C

QUESTION   ANSWER

116. For what percentage of March was it raining?

   a.   18%
   b.   21%
   c.   26%
   d.   29%
   e.   42%

QUESTION   ANSWER

117. What percentage of the total number of rain
      days (in the six months presented) occurred in
      November?

      a.   14%
      b.   16%
      c.   18%
      d.   20%
      e.   22%

QUESTION   ANSWER

118. Between which two months was the smallest
      percentage decrease in the number of rain days?

      a.   January & March
      b.   March & May
      c.   May & July
      d.   July & September
      e.   September & November

QUESTION   ANSWER

119. If the mean temperature from May to August
      increased by the same percentage as it increased
      from March to May, what was the mean
      temperature in August?

      a.   28°C
      b.   30°C
      c.   31°C
      d.   32°C
      e.   34°C

QUESTION	ANSWER

120. If the mean temperature from September to October decreased by 10%, what was the percentage decrease from October to November?

    a.   21.8%

    b.   23.7%

    c.   25.9%

    d.   27.2%

    e.   29.5%

QUESTION	ANSWER

121. If the total number of rain days for the year was 108, what percentage of the rain days was in the first quarter of the year?

    a.   15%

    b.   16%

    c.   17%

    d.   18% or above

    e.   Cannot say

If you feel comfortable with percentages let's move on to ratios. If you are not comfortable with a particular rule, it is better to go back and read through that rule again before you move on.

Ratios

Ratios are used to make comparisons between two (or more) quantities. The most common way to present ratios is by writing 'the ratio of A to B', or by writing 'A:B'. If the ratio of A to B is 2:1, this simply means that A is two times greater than B.

Note that 2:1 is the same as 4:2, or 6:3, or 3:1.5, or 0.5:0.25. In theory there are an infinite number of ways in which you can express the relationship between two quantities as a ratio. This is because as long as you multiply (or divide) both sides of the ratio with the same number, the ratio is stating the same relationship: that A is two times greater than B.

The clearest way to present this relationship is to convert the ratio to X:1, because this can be read as A is X times greater than B. However, you should also be familiar with ratios in their less simplified form (i.e. as 6:3 rather than 2:1), as they are often presented as such in multiple-choice answers.

One of the most common questions you can be asked is to express the ratio of two quantities.

GENERAL QUESTION: What is the ratio of X to Y, if X = a and Y = b?
GENERAL SOLUTION: Divide a by b and then write the value ':1'.
FORMULA: $(a \div b):1$

EXAMPLE QUESTION: If Steve used 8 medium boxes and 14 small boxes for packing, what was the ratio of medium to small boxes?

EXAMPLE SOLUTION: You can express two quantities as a ratio by dividing the first quantity by the second. In this example you need to divide 8 by 14, which gives you 0.5714. This number tells you that the first quantity (medium boxes) is 0.5714 times the second quantity. To turn this into a ratio you just need to write it as 0.5714:1. Or, if you round this off to two decimals, it is 0.57:1.

TIP: Note that here we divided 8 by 14 because we were requested to give the ratio of medium to small boxes. If we were asked for the ratio of small to medium boxes, we would have divided 14 by 8 instead. This would have given us a ratio of 1.75:1.

Practice questions

QUESTION	ANSWER

122. What is the ratio of X to Y if X = 130 and Y = 170?

    a.   0.76:1
    b.   0.79:1
    c.   0.81:1
    d.   0.87:1
    e.   1.31:1

QUESTION	ANSWER

123. What is the ratio of X:Y if X = 0.15 and Y = 0.45?

    a.   0.30:1
    b.   0.33:1
    c.   0.35:1
    d.   0.37:1
    c.   0.40:1

QUESTION	ANSWER

124. What is the ratio of X:Y if X is 20 and Y is 14?

    a.   0.70:1
    b.   1.14:1
    c.   1.27:1

d.   1.35:1

e.   1.43:1

QUESTION   ANSWER

125. What is the ratio of male to female students in a
      classroom, if there are 22 males and 16 females?

    a.   0.73:1

    b.   0.75:1

    c.   1.35:1

    d.   1.38:1

    e.   1.40:1

QUESTION   ANSWER

126. In the same classroom (as in the previous ques-
      tion), what is the ratio of females to males?

    a.   0.73:1

    b.   0.75:1

    c.   1.35:1

    d.   1.38:1

    e.   1.40:1

QUESTION   ANSWER

127. A recipe for a cake requires half a cup of sugar
      and 2 cups of milk. What would be the ratio of
      sugar to milk in this cake?

    a.   0.25:1

    b.   0.30:1

    c.   0.5:1

d.   0.75:1

e.   4:1

FIGURE 4: **Distribution of labour force**

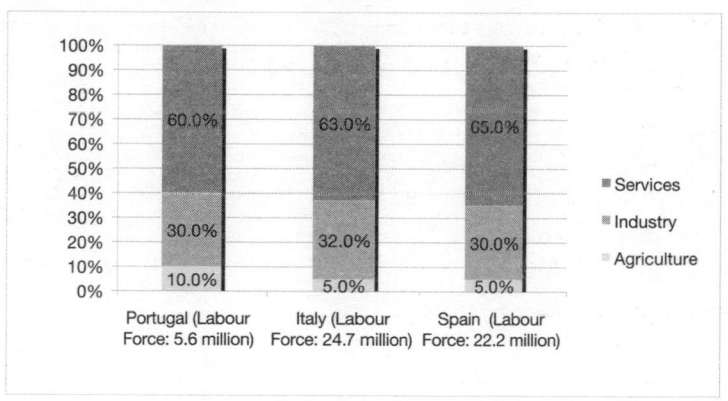

QUESTION   ANSWER

128.   What is the ratio of people working in industry
       to people working in agriculture in Italy?

   a.   3:1

   b.   5.7:1

   c.   6:1

   d.   6.4:1

   e.   7:1

QUESTION   ANSWER

129.   What is the ratio of people working in services
       in Portugal to people working in services in
       Spain?

a.   0.23:1
b.   0.27:1
c.   0.32:1
d.   0.41:1
e.   0.92:1

QUESTION   **ANSWER**

130. What is the ratio of the labour force in Italy to
      that in Portugal?

a.   0.23:1
b.   3.8:1
c.   4.4:1
d.   4.6:1
e.   5.1:1

QUESTION   **ANSWER**

131. What is the ratio of people working in services
      in Portugal to people working in agriculture in
      Spain?

a.   0.33:1
b.   2.85:1
c.   2.92:1
d.   2.98:1
e.   3.03:1

QUESTION   **ANSWER**

132. In Portugal, what is the ratio of people working
      in services to people working in industry to
      people working in agriculture?

a.   1:3:6
b.   3:6:1
c.   6:1:3
d.   6:3:1
e.   5.6:3:1

QUESTION   ANSWER

133. In Portugal, what is the ratio of people working
in services to people working in agriculture to
people working in industry?

a.   2:0.33:1
b.   2:0.5:1
c.   3:0.5:1
d.   3.2:2:1
e.   3.5:0.7:1

QUESTION   ANSWER

134. In Italy, what is the ratio of people working in
services to people working in industry to people
working in agriculture?

a.   10.8:4.1:1
b.   12.6:6.4:1
c.   15.5:7.5:1
d.   20:12:1
e.   30:15:1

There is another type of question that you might be asked on ratios:
you might be given the ratio of two quantities and the size of one of
the two, and then asked to calculate the size of other quantity.

GENERAL QUESTION: If the ratio of X to Y is a:b, and X = x, what is the size of Y?

GENERAL SOLUTION: Multiply x by b divided by a.

FORMULA: $x \times (b \div a)$

EXAMPLE QUESTION: If the ratio of black to red pens in a store is 1.2:1, and there are 312 black pens, how many red pens are there?

EXAMPLE SOLUTION: You can calculate the number of red pens by multiplying 312 by (1 ÷ 1.2), which equals 312 ÷ 1.2. This will give you the solution of 260 red pens.

You might be given the size of the second group, and asked to calculate the size of the first quantity.

GENERAL QUESTION: If the ratio of X to Y is a:b, and Y = y, what is the size of X?

GENERAL SOLUTION: Multiply y by a divided by b.

FORMULA: $y \times (a \div b)$

EXAMPLE QUESTION: If the ratio of black to red pens in a store is 1.2:1, and there are 260 red pens, how many black pens are there?

EXAMPLE SOLUTION: You can calculate the number of black pens by multiplying 260 by (1.2 ÷ 1), which equals 260 × 1.2. This will give you the solution of 312 black pens.

Practice questions

QUESTION   ANSWER

135. If the ratio of X to Y is 5:1, and X is 140,
what is Y?

    a.   20
    b.   24
    c.   25
    d.   28
    e.   32

QUESTION   ANSWER

136. If the ratio of X to Y is 2.5:2, and X is 35,
what is Y?

    a.   14
    b.   18
    c.   28
    d.   30
    e.   31

QUESTION   ANSWER

137. If the ratio of X to Y is 4:1, and Y is 2,000,
what is X?

    a.   6,000
    b.   8,000
    c.   8,500
    d.   9,000
    c.   10,000

QUESTION   ANSWER

138. If the ratio of X to Y is 3:2, and Y is 40, what is X?

    a.   45

    b.   50

    c.   55

    d.   60

    e.   65

QUESTION   ANSWER

139. If the ratio of trousers to skirts in a shop is 9:2 and there are 585 pairs of trousers, how many skirts are there?

    a.   130

    b.   135

    c.   140

    d.   145

    e.   150

QUESTION   ANSWER

140. If the ratio of shirts to T-shirts in a shop is 3:5 and there are 350 T-shirts, how many shirts are there?

    a.   185

    b.   190

    c.   195

    d.   200

    e.   210

Another type of question that you can be asked on ratios, is when you are given the total size of a group (T), and the ratio of its subgroups, and you are asked to calculate the size of one of the subgroups.

GENERAL QUESTION: If T consists of A and B, and the ratio of A:B is x:y, what is the size of A?

GENERAL SOLUTION: Divide x by the sum of x plus y, and multiply this by the total size of the group (T).

FORMULA: $T \times (x \div (x + y))$

EXAMPLE QUESTION: If there are 350 people working in a firm, with a male to female ratio of 4:3, how many males are in the firm?

EXAMPLE SOLUTION: The problem tells you that there are 4 males for every 3 females in the firm. This means in every 7 workers (4 + 3), there are 4 males, i.e. 4 out of 7 workers are males. So to find the number of males in 350 workers, you need to calculate $350 \times (4 \div 7) = 200$.

Or, the question might ask you to calculate the size of the other subgroup.

GENERAL QUESTION: If T consists of A and B, and the ratio of A:B is x:y, what is the size of B?

GENERAL SOLUTION: Divide y by the sum of x plus y, and multiply this by the total size of the group (T).

FORMULA: $T \times (y \div (x + y))$

Let's see this as a variation of the previous example:

> **EXAMPLE QUESTION:** If there are 350 people working in a firm, with a male to female ratio of 4:3, how many *females* are in the firm?
>
> **EXAMPLE SOLUTION:** As we saw previously, there are 3 females in every 7 employees. In order to find the number of females in a total of 350 employees, you need to calculate $350 \times (3 \div 7)$. The solution is 150 females.

Practice questions

QUESTION	ANSWER

141. If T = 700, and T consists of A and B, with a ratio of 3:2, what is the size of A?

    a.   280
    b.   380
    c.   390
    d.   400
    e.   420

QUESTION	ANSWER

142. If T = 700, and T consists of A and B, with a ratio of 4:1, what is the size of B?

    a.   100
    b.   140
    c.   160
    d.   200
    e.   560

QUESTION   ANSWER

143. If T = 3,000, and T consists of A and B, with a
ratio of 0.5:1, what is the size of A?

a.   500
b.   1,000
c.   1,200
d.   1,500
e.   3,000

QUESTION   ANSWER

144. If T = 3,000, and T consists of A and B, with a
ratio of 8:7, what is the size of B?

a.   1,200
b.   1,350
c.   1,400
d.   1,550
e.   1,600

QUESTION   ANSWER

145. If a class has 120 students, and the ratio of students
passing all their exams is 9:1, how many people
failed at least 1 exam?

a.   9
b.   10
c.   11
d.   12
e.   15

QUESTION   ANSWER

146. If a class has 140 students, and the ratio of males
to females is 4:3, how many males are in the class?

    a.   60

    b.   74

    c.   76

    d.   78

    e.   80

Combining averages, ratios and percentages

Now that you have practised basic calculations, averages, ratios and
percentages, you need to practise with various combinations of these
types of questions. It is easier to work through any question after you
have seen its answer, or after you have solved several similar questions.
But when you are taking a numerical reasoning test, the questions will
be mixed, so it's important to practise a variety of questions, without
being told which type of method you should use to solve it.

FIGURE 5: **Natural gas (in million cubic metres)**

	Germany	Netherlands	Italy	France	UK
Production	21,000	75,000	18,000	2,000	104,000
Consumption	94,000	50,000	73,000	43,000	94,000
Exports	7,000	46,000	1,000	1,000	15,000
Imports	80,000	?	?	42,000	5,000

QUESTION   **ANSWER**

147. What is the average consumption of natural gas of
the five countries presented, in billion cubic metres?

    a.   44

    b.   53.5

    c.   70.8

    d.   53,500

    e.   70,800

TIP: Always pay attention to the number of zeros and decimals. In this
example, if you didn't notice that the table is in million cubic metres
and the question in billion cubic metres, you would have chosen
the wrong answer.

QUESTION   **ANSWER**

148. Approximately what percentage of the natural gas
produced and imported in Germany is consumed?

    a.   88%

    b.   90%

    c.   91%

    d.   93%

    e.   96%

QUESTION   **ANSWER**

149. How much more natural gas is produced in
Germany than in Italy?

    a.   17%

    b.   19%

    c.   20%

    d.   24%

    e.   29%

TIP: It often helps to have a quick peek at the answer options before making any calculations. For example, in this question the answer could have been an integer instead of a percentage (i.e. 21,000 − 18,000 = 3,000 million m³ more). If both answers are legitimate, glancing at the options will prevent your wasting time making unnecessary calculations.

....

QUESTION   **ANSWER**

150. What is the proportion of the natural gas produced to that consumed in the UK?

    a.   1.05:1

    b.   1.11:1

    c.   1.13:1

    d.   1.17:1

    e.   1.19:1

....

QUESTION   **ANSWER**

151. If gas can either be consumed or exported, how much gas was imported in the Netherlands?

    a.   17,000

    b.   18,000

    c.   19,000

    d.   20,000

    e.   21,000

QUESTION   **ANSWER**

152. If gas can either be consumed or exported, what
is approximately the ratio of the imports of natural
gas in Italy to the imports in France?

    a.   1.12:1

    b.   1.26:1

    c.   1.33:1

    d.   1.47:1

    e.   1.51:1

FIGURE 6: **Age structure in the UK and France**

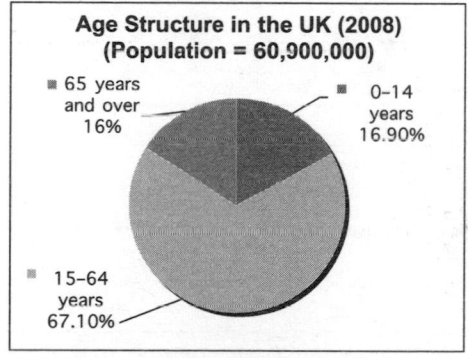

Age Structure in the UK (2008)
(Population = 60,900,000)

65 years and over 16%

0–14 years 16.90%

15–64 years 67.10%

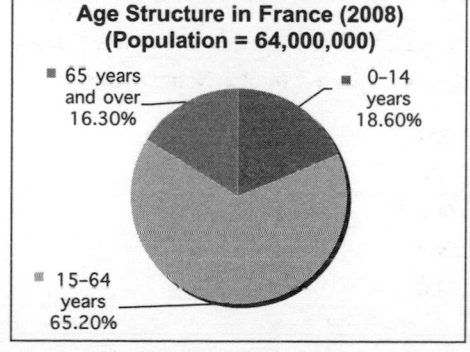

Age Structure in France (2008)
(Population = 64,000,000)

65 years and over 16.30%

0–14 years 18.60%

15–64 years 65.20%

QUESTION	ANSWER

153. How many people in France were aged 65 years and over in 2008?

    a.   9,744,000
    b.   9,850,000
    c.   9,978,000
    d.   10,432,000
    e.   11,904,000

QUESTION	ANSWER

154. How many people in the UK were aged 15 years and over in 2008?

    a.   50,208,000
    b.   50,302,600
    c.   50,607,900
    d.   51,270,000
    e.   52,160,000

QUESTION	ANSWER

155. What percentage of the UK population was aged less than 65 years in 2008?

    a.   83%
    b.   83.8%
    c.   84 %
    d.   84.2%
    e.   85%

QUESTION   ANSWER

156. How many people were aged 15–64 years in France,
in comparison to the UK?

    a.   688,000 more

    b.   688,000 fewer

    c.   864,100 more

    d.   864,100 fewer

    e.   1,611,900 more

QUESTION   ANSWER

157. What was the ratio of people aged 14 years or less
in the UK to those in France in 2008?

    a.   0.86:1

    b.   0.89:1

    c.   0.91:1

    d.   0.96:1

    e.   1.16:1

QUESTION   ANSWER

158. If the population in the UK increased by 4% from
2000 to 2008, what was the UK population in 2000?

    a.   58,464,000

    b.   58,557,692

    c.   58,598,504

    d.   58,616,822

    e.   58,620,050

QUESTION	ANSWER

159. If in 2008, 51.2% of the people aged less than 15 years in France were male, how many was this?

    a.   5,269,555

    b.   5,341,184

    c.   5,825,240

    d.   5,890,520

    e.   6,094,848

QUESTION	ANSWER

160. If in 2008, 44% of the people aged 65 years and over in the UK were men, what was the male to female ratio in that age group?

    a.   0.76:1

    b.   0.79:1

    c.   0.82:1

    d.   0.88:1

    e.   1.28:1

QUESTION	ANSWER

161. If there are 30,770,000 women in the UK, and the male to female ratio over the age of 64 is 0.8:1, how many women are aged 64 years or less?

    a.   5,413,333

    b.   24,377,252

    c.   25,356,667

    d.   25,799,333

    e.   Cannot say

Unit conversions

You may also face questions about conversions between different units when sitting a numerical reasoning test. These can involve converting units of length, volume, weight, temperature, currency and so on.

Such questions are actually linked to ratios. You will be given the relationship of the first unit to the second, which is essentially their ratio, even if this is not made explicit in the question.

Let's look at some examples:

162. How many kilos are equal to 45 pounds?
      (1 pound = 0.4536 kilos)

    a.   16.8
    b.   18.2
    c.   19.5
    d.   20.2
    e.   20.4

Here you are given the equivalent number of kilos for one pound, and are asked to convert a number of pounds into kilos. The ratio of kilos to pounds is 0.4536:1. So all you need to do is multiply the number of pounds (45) by the equivalent kilos to a pound: 45 × 0.4536 = 20.4 kilos.

163. How many pounds are equal to 300 kilos?
      (1 pound = 0.4536 kilos)

    a.   661.4
    b.   670.2
    c.   673.6
    d.   675.1
    e.   678.4

Here you are again given the equivalent kilos for one pound, but this time you are asked to convert a number of kilos into pounds. So you need to perform the opposite type of calculation. You need to divide 300 by 0.4536 to find how many pounds 300 kilos are: $300 \div 0.4536 = 661.4$ pounds.

TIP: In unit conversions you can sanity-check your answer (to ensure that you haven't confused multiplication with division) by checking that there is a greater number of the smaller unit, or vice versa. Here, for example, it is clear that kilos are larger than pounds, so there must be more pounds than kilos. Your answer, therefore, should definitely be greater than 300. If you had used multiplication instead of division, you could immediately spot that you had made a mistake, because your answer would have been $300 \times 0.4536 = 136.1$.

Practice questions

QUESTION	ANSWER

164. How many gallons are equal to 30 litres?
(1 litre = 0.2642 gallons)

    a.   6.2
    b.   6.8
    c.   7.9
    d.   8.1
    e.   9.2

QUESTION	ANSWER

165. How many litres are equal to 1,500 gallons?
(1 litre = 0.2642 gallons)

a.   396.3
b.   5592.8
c.   5614.6
d.   5677.5
e.   5693.4

QUESTION   ANSWER

166. How many miles are equal to 250 kilometres?
(100 km = 62.1371 miles)

a.   152.5
b.   155.3
c.   157.1
d.   158.5
e.   159.4

TIP: Note that here you were given the equivalent number of miles for
100 km, not 1 km. This added an extra step to your calculations.
You had to multiply 250 by 62.1371, and then divide this by 100.

QUESTION   ANSWER

167. How many kilometres are equal to 400 miles?
(10 km = 6.2137 miles)
a.   630.2
b.   635.5
c.   639.1
d.   641.4
e.   643.7

QUESTION   ANSWER

168. How many ounces are equal to 20 kilos?
(5 kilos = 160.75 ounces)

a.   630
b.   637
c.   643
d.   649
e.   651

QUESTION   ANSWER

169. How many kilos are equal to 700 ounces?
(5 kilos = 160.75 ounces)

a.   21.77
b.   23.65
c.   24.31
d.   26.23
e.   27.15

FIGURE 7: **Currency conversions for UK pounds (GBP)**

	US Dollars (USD)	Japanese Yen (JPY)	Canadian Dollars (CAD)	Euro (EUR)	Australian Dollars (AUD)	Indian Rupees (INR)
1 GBP is	1.4774	133.554	1.79201	1.05532	2.16369	70.5163

QUESTION	ANSWER

170. How many Japanese yen are equal to 380 UK pounds?

    a.   50,750.52 JPY
    b.   50,820.24 JPY
    c.   50,870.22 JPY
    d.   50,910.15 JPY
    e.   50,960.50 JPY

QUESTION	ANSWER

171. How many US dollars are equal to 2,000 UK pounds?

    a.   2,880.25 USD
    b.   2,895.20 USD
    c.   2,920.55 USD
    d.   2,954.80 USD
    e.   3,010.75 USD

QUESTION	ANSWER

172. How many UK pounds do you need to buy 1,500 Australian dollars?

    a.   688.23 GBP
    b.   693.26 GBP
    c.   695.37 GBP
    d.   698.12 GBP
    e.   699.46 GBP

QUESTION   ANSWER

173. How many UK pounds do you need to buy 20,000
     Canadian dollars?

     a.   11,100.15 GBP
     b.   11,125.25 GBP
     c.   11,160.65 GBP
     d.   11,172.35 GBP
     e.   11,194.55 GBP

QUESTION   ANSWER

174. What are 500 euros worth in Indian rupees?

     a.   29,197.52 INR
     b.   29,238.13 INR
     c.   30,432.56 INR
     d.   32,502.11 INR
     e.   33,409.91 INR

QUESTION   ANSWER

175. What are 120 Australian dollars worth in Japanese
     yen?

     a.   7,388.74 JPY
     b.   7,391.52 JPY
     c.   7,401.93 JPY
     d.   7,403.17 JPY
     e.   7,407.01 JPY

QUESTION   ANSWER

176.   What are 5,000 Indian rupees worth in Canadian
           dollars?

   a.   127.06 CAD
   b.   130.48 CAD
   c.   131.29 CAD
   d.   131.93 CAD
   e.   133.19 CAD

Sequences

Questions on sequences can follow any rule that a psychometrician can
come up with, which makes it impossible to cover every single one
here. Having said that, there are some standard rules that are very
frequently used in sequence questions.

The trick with sequences, however, is not to memorise the rules. It
is to solve as many sequence questions as possible, so that you can
quickly recognise a rule that you have already seen before and also get
used to coming up with new solutions, so that you can crack the puzzle
if it is one you have not yet come across.

Here we will help you develop both of these skills. We will take you
through a few rules that you are likely to encounter and will also give
you sequences with rules that we haven't described, so that you can
practise getting to the solution from scratch.

The first thing you have to do when you see a sequence is observe
the numbers and the relationship between them. You usually start from
the beginning and work through the numbers in pairs, as typically each
number is related to those adjacent to it. Psychometricians may occasionally link non-adjacent numbers to make the question more complex,
but this would be the exception rather than the rule.

So let's suppose you have the following sequence: 15, 18, 21, 24, ? . . .

You could start by thinking, what is the relationship between 15 and 18? There are infinite ways in which two numbers can be related, so you would typically narrow these down to a few that are most obvious. For example:

1. the second number equals the first number 'plus 3',
2. the first number is 3×5 and the second is 3×6,
3. the first number is $16 - 1$ and the second is $16 + 2$,
4. the first number is 7.5×2 and the second is 9×2.

Once you have a few 'rules', you move to the next pair. What is the relationship between 18 and 21? Now you go through the same 'rules' that you found in the first pair, and see if any of these also stand for the second pair:

1. the second number equals the first number 'plus 3',
2. the first number is 3×6 and the second is 3×7,
3. the first number is $16 + 2$ and the second is $16 + 5$,
4. the first number is 9×2 and the second is 10.5×2.

It is easy to see that your first rule is correct. Once you have a rule that is correct for two pairs, you can quickly double-check it with the other pairs. Every number equals its previous number in the sequence plus 3. So you can easily use this rule to find that the next number in the sequence would be $24 + 3 = 27$.

Note that in this example your second rule was correct as well, and it would have led you to the same answer. When you have found a rule that works for all the numbers that you are given, there is no need to check any of the other rules. Even if other rules are also correct (which is only the case when they are essentially saying the same thing), you don't need to waste time trialling them.

RULE: **Adding (or subtracting, or multiplying or dividing by) a constant number**

One of the most common rules behind a sequence is the addition of a constant number. This rule can be used in an infinite number of ways, as it can start from any number sequence, and the number which is added can also have any value. How does this help you? You can very quickly check and discredit this rule, by subtracting the first two pairs of numbers. If you get the same answer, you can check the next pairs.

Have a go at the following example.

177. Find the number that should replace '?' in the following sequence: −31, −19, −7, 5, 17, ? . . .

    a.   −29
    b.   21
    c.   26
    d.   29
    e.   31

The difference between the first two numbers, is −19 − (−31) = −19 + 31 = 12. If you then check the difference between the second and third numbers, this is −7 − (−19) = −7 + 19 = 12. You will find the same with the next pairs as well. So the rule behind this sequence is adding 12 to each consecutive number. If you apply this to the last number of the sequence, you will find that the solution is 17 + 12 = 29.

Practice questions

Find the number that should replace '?' in the following sequences.

QUESTION   ANSWER

178. 2, 6, 10, 14, 18, ? . . .

    a.   20

    b.   22

    c.   24

    d.   26

    e.   28

QUESTION   ANSWER

179. 132, 147, 162, 177, 192, ? . . .

    a.   202

    b.   205

    c.   207

    d.   208

    e.   211

QUESTION   ANSWER

180. −10, −6, −2, 2, ? . . .

    a.   −10

    b.   −4

    c.   4

    d.   6

    e.   10

A variation of this rule is, instead of adding, to subtract, or multiply or divide by a constant number.

181. Find the number that should replace '?' in the following sequence: 35, 31, 27, 23, 19, ? . . .

    a.   12

    b.   13

    c.   14

    d.   15

    e.   16

If you check the difference between the first two numbers, this is 31 − 35 = −4. If you then check the difference between the second and third numbers, this is 27 − 31 = −4. You will find the same with the following pairs as well. So the rule behind this sequence is subtracting 4 from each consecutive number. If you apply this to the last number of the sequence, you will find that the solution is 19 − 4 = 15.

Practice questions

Find the number that should replace '?' in the following sequences.

....

QUESTION   ANSWER

182. 88, 81, 74, 67, ? . . .

    a.   64

    b.   61

    c.   60

    d.   58

    e.   57

QUESTION   **ANSWER**

183. 1, 2, 4, 8, 16, ? . . .

    a.   32

    b.   34

    c.   36

    d.   38

    e.   40

QUESTION   **ANSWER**

184. 277, 265, 253, 241, ? . . .

    a.   239

    b.   237

    c.   235

    d.   233

    e.   229

QUESTION   **ANSWER**

185. ⅓, 1, 3, 9, 27, ? . . .

    a.   36

    b.   54

    c.   63

    d.   81

    e.   243

QUESTION   **ANSWER**

186. 25, 100, 400, 1,600, ? . . .

    a.   2,000

    b.   2,400

c.   3,200
d.   5,000
e.   6,400

QUESTION   ANSWER

187.   3, 6, 12, 24, ?, 96 . . .

a.   44
b.   48
c.   52
d.   54
e.   56

QUESTION   ANSWER

188.   61, 62, 60, 63, 59, ? . . .

a.   58
b.   62
c.   64
d.   65
e.   66

QUESTION   ANSWER

189.   −3, −7, −11, −15, −19, ? . . .

a.   −23
b.   −24
c.   −25
d.   −26
e.   −27

RULE: Adding (or subtracting, or multiplying or dividing by) a non-constant number

Now let's look at how sequences can become more complicated. Instead of following the basic calculations with a constant number, the same rule of adding (or subtracting, or multiplying or dividing by) a number can also be used with non-constant numbers: numbers that themselves follow a sequence. You could therefore be looking for a sequence within a sequence! It is not as complicated as it sounds though.

As we have already seen, the most common rule that you can look for is one where numbers are either added, subtracted, multiplied or divided. The more examples you solve, the easier it will be to identify them. And the easier it will be to identify them within another sequence. Let's look at an example.

190. Find the number that should replace '?' in the
      following sequence: 5, 7, 11, 17, 25, ? . . .

    a.   10
    b.   28
    c.   30
    d.   33
    e.   35

If you check the difference between the first two numbers, this is 7 − 5 = 2. If you then check the difference between the second and third numbers, this is 11 − 7 = 4. The difference of the next pair is 17 − 11 = 6, and the difference of the final pair is 25 − 17 = 8. So in the main sequence, instead of adding a constant number, you are adding a number that is increasing by 2 each time. So far you have added the numbers 2, 4, 6 and 8. The next number of this 'inner' sequence is 8 + 2, which equals 10. So you have to add 10 to 25 to get the next number of the main sequence, and your solution will be option **e**: 35.

Practice questions

Find the number that should replace '?' in the following sequences.

....

QUESTION	ANSWER

191. 2, 4, 7, ?, 16, 22 . . .

a. 9
b. 11
c. 12
d. 15
e. 29

....

QUESTION	ANSWER

192. 4, 8, 24, 96, 480, ? . . .

a. 640
b. 960
c. 1,240
d. 2,880
e. 3,600

....

QUESTION	ANSWER

193. 2, 6, 14, 30, ?, 126 . . .

a. 59
b. 60
c. 62
d. 66
e. 68

....

QUESTION	ANSWER

194. 50, 48, 44, ?, 30 . . .

a.   32
b.   34
c.   36
d.   38
e.   40

QUESTION   ANSWER

195.   3, 5, 9, 15, 23, ?, 45 . . .

a.   30
b.   32
c.   33
d.   38
e.   40

QUESTION   ANSWER

196.   61, 62, 60, 63, 59, ? . . .

a.   57
b.   60
c.   63
d.   64
e.   65

QUESTION   ANSWER

197.   3, 2, 4, 1, 5, ? . . .

a.   −1
b.   0
c.   1
d.   2
e.   4

RULE: **Powers and roots**

Another rule that sequences frequently follow involves powers (exponents) or roots. When each number of the sequence equals the previous one to a power, this is typically easy to spot, because the difference between the numbers increases very rapidly.

198. Find the number that should replace '?' in the following
     sequence: 2, 8, 512, 134,217,728, ? . . .

    a. 268,435,456

    b. 328,900,572,122

    c. 2,048,289,502,155

    d. 18,014,398,509,481,984

    e. 2,417,851,639,229,258,349,412,352

This exponential increase between consecutive numbers is an indication that powers are involved, so you can save time by trialling relationships between the numbers that involve powers (rather than additions or subtractions etc.). Each number equals the previous number to the power of 3. The next number in this sequence would be $134,217,728^3$, which equals 2,417,851,639,229,258,349,412,352.

Another way in which powers and roots are used in sequences is when there is an existing sequence in the numbers, and these numbers are raised to the same power.

199. Find the number that should replace '?' in the following sequence: 4, 36, 100, 196, 324, ? . . .

    a.   361

    b.   400

    c.   441

    d.   484

    e.   529

All the numbers of this sequence are numbers raised to the second power: $2^2, 6^2, 10^2, 14^2, 18^2$. Also, the numbers 2, 6, 10, 14 and 18 are numbers that consecutively increase by 4 units. So the following number in this sequence would be $(18 + 4)^2 = 22^2$, which equals 484.

Practice questions

Find the number that should replace '?' in the following sequences.

....

    QUESTION   **ANSWER**

200.   1, 4, 9, 16, 25, ? . . .

    a.   30

    b.   35

    c.   36

    d.   40

    e.   49

....

    QUESTION   **ANSWER**

201.   196, 169, 144, 121, 100, ? . . .

    a.   64

    b.   81

    c.   89

d.   91
e.   99

QUESTION   ANSWER

202.   8, 27, 64, 125, ?, 343 . . .

a.   136
b.   144
c.   200
d.   216
e.   250

QUESTION   ANSWER

203.   4, 16, 256, 66,536, ? . . .

a.   262,144
b.   1,048,576
c.   16,777,216
d.   166,264,116
e.   4,294,967,296

Overview of generalisations

Here are all the generalisations mentioned in this chapter so that you can quickly glance through them while practising.

Rounding off numbers

Digits from 0–4 are rounded down. Digits from 5–9 are rounded up. When you are rounding off, you only take into account the first digit on the right.

Averages

What is the average of X, Y and Z?

$(X + Y + Z) \div N$ (where N is the number of values)

If the average of X, Y and Z is A, what is the value of X, if you know the values of Y, Z and A?

$(A \times$ number of values$) - Y - Z$

Percentages

What is A as a percentage of C?

$(A \div C) \times 100\%$

What is x% of A?

$A \times (x \div 100)$

What is A increased by x%?

$(A \times x\%) + A$, *or:* $A \times (1 + x\%)$

What was B before it increased by x%?

$B \div (1 + x\%)$, *or:* $B \div ((100 + x) \div 100)$

What is A decreased by x%?

$A \times (1 - x\%)$, *or:* $A \times ((100 - x) \div 100)$

Ratios

What is the ratio of X to Y, if X = a and Y = b?

$(a \div b){:}1$

What is the ratio of Y to X, if X = a and Y = b?

$(b \div a){:}1$

If the ratio of X to Y is a:b, and X = x, what is the size of Y?
x × (b ÷ a)

If the ratio of X to Y is a:b, and Y = y, what is the size of X?
y × (a ÷ b)

If T consists of A and B, and the ratio of A:B is x:y, what is the size of A?
T × (x ÷ (x + y))

If T consists of A and B, and the ratio of A:B is x:y, what is the size of B?
T × (y ÷ (x + y))

Summary and next steps

You have now gone through the most common types of questions asked in numerical reasoning tests. You have seen the logic behind their solutions, you have seen some 'tricks' that psychometricians use to turn simple questions into more complex ones and you have learned a few shortcuts that can save you some time when sitting a test. You have also had a chance to practice over 180 numerical reasoning questions, with data presented in different kinds of formats, including tables, bar charts, pie charts and graphs. Finally, you have had some practice at bringing together the different types of questions and all that you have learned throughout this chapter.

Do you feel ready? If you are not sure, take a break for a day or two (if your deadline for taking the test allows you to do so), and then have a go at the test presented in Chapter 5. Evaluate your performance. If you need further practice, you can use *Perfect Numerical Test Results*, which includes more numerical reasoning questions for you to try, or another book that focuses on numerical reasoning tests. There are also several websites where you can access online tests (see page 180 for details). This will be especially useful if you have to take the test online, as it can also help you become familiar with the format of the test.

Step 4: Master your opponent – logical reasoning tests

In logical reasoning tests, also known as abstract, diagrammatic or non-verbal tests, you are asked to solve visual puzzles, a task that eliminates any cultural differences, language difficulties or mathematical inadequacies! The logic behind these tests is simple: find the underlying rules between shapes, patterns or objects. Once the sequence is clear to you, you will be able to identify the next logical figure from the number of options given. If this sounds complicated, don't worry: you will find it quite simple once you have read the 'basic rules' section later in this chapter (see page 122).

The questions or items in logical reasoning tests can vary in difficulty. Most tests will start with simple figures so that you have a clear picture of what you are supposed to do. As you advance through the test, the figures will become more complicated and the rules that underlie them, harder to identify. However, if you cannot find the logic behind one question, it does not mean you will be unable to work out the others. You could, if necessary, skip the one you're struggling with and move to the next question.

In order to understand the logical reasoning questions, first you need to become familiar with the types of questions that are typically used, and the forms in which they are likely to appear.

Types of logical reasoning questions

Which figure completes the diagram?

Five in a row

This is the most common form of logical reasoning test question. You are given a number of boxes, usually four or five, one adjacent to another, with shapes or objects in each box. The last box contains a question mark, which means that you need to fill it in with an object or shape that is a logical continuation of the contents of the previous boxes.

Beneath the boxes you will typically find four or five possible answers, marked with letters (A, B, C, etc.), from which you have to choose the one that logically follows the sequence above. An example is given below. The logic behind it, which explains the correct answer, is given after the question, but have a go at solving it on your own first.

1.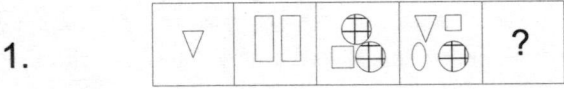

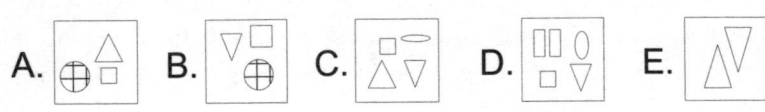

In the five boxes on top, the first box contains only one shape, the second contains two shapes, the third one three and the last one four shapes. The basic rule that underlies this sequence is that each box contains one shape more than the number of shapes in the previous box. Since the last box contains four shapes, the next in line must contain five shapes. Looking at the five possible answers beneath, you can see that boxes A and B contain three shapes each, box C contains four shapes, box D has five shapes and box E has two shapes. Therefore the correct answer is box D.

Variations of this type of question

One variation of the five-in-a-row question is when the question mark is not located in the last box but in one of the others, as shown in the two examples below. In this variation you are asked to pick the answer that fills the box with the question mark to produce a logical sequence.

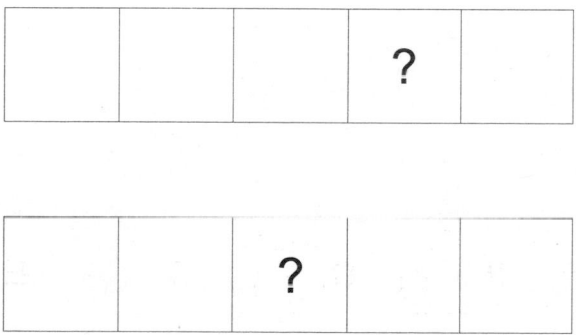

Another variation is when there aren't five boxes, as there are in the examples above, but four boxes, or even six.

2×2 table

A different way of presenting the data would be in four boxes forming a square. In this case there are shapes in three of the boxes and you are asked to fill the fourth one by choosing one of the boxes presented below.

2.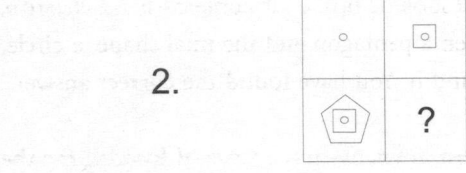

As you can see, the first box (on the top left) contains one circle. The second box (top right) contains the same circle, surrounded by a square. The third (bottom left) contains the two previous shapes, surrounded by a pentagon. Therefore, the rule behind this example is that each box contains the contents of the previous box, surrounded by another shape.

One way of finding the correct answer, from the five given beneath the question, is to look for the shapes you have been given in the third box and see how many possible answers contain those exact shapes, surrounded by another. You can only find this in box C.

Another way of locating the correct answer is by the method of elimination. When looking at the five possible answers you can quickly eliminate box D because it does not contain the circle in the middle. You can also eliminate box B since it contains exactly the same shapes as the second box in the question. Box E can also be eliminated because it has a triangle around the small circle, which does not follow the circle-square-pentagon sequence in the question. The same applies to box A, since it contains the same triangle. This means that box C is the only remaining possibility.

You must always check that the remaining box contains the correct answer, because there is a chance that you have misinterpreted the rule underlying the item. If you look at box C, it contains a small circle, surrounded by a square, then a pentagon and the final shape, a circle, which has been added around it. You have found the correct answer.

When given a 2×2 table item, there are three ways of looking for the logical sequence:

1. You can look for the logic by starting with the upper left box, followed by the upper right box and then continuing with the lower boxes, first the left and then the right one (which is the way we looked at the previous example).

1	2
3	4

2. Or you can start with the upper left box, then move to the lower left box, followed by the upper right box, and finally the lower right box, as shown on the next page. This is not very common, but is useful to know just in case.

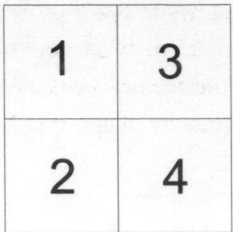

Here is an example using this alternative way of looking at the table.

3.

A. B. C. D. E.

In this example, you can see in the first box (upper left) one single line. The second box (lower left) has an 'L' shape with two sides. The third (upper right) has a shape with three sides. The rule that applies to this example is that as the boxes advance, the shapes in them have one more side than the shape in the previous box. From the five possible answers given to you, you are looking for a shape that has four sides. The only one with four sides is in box B. The shape in box A has two sides, box C has five sides, box D has three sides and box E has eight sides.

3. Another rule that might apply to a 2×2 table item is when it is shown as two pairs. The two upper boxes are related in a way that will help you find the link between the two lower boxes and, consequently, the missing shape for the box. Have a go at the following example.

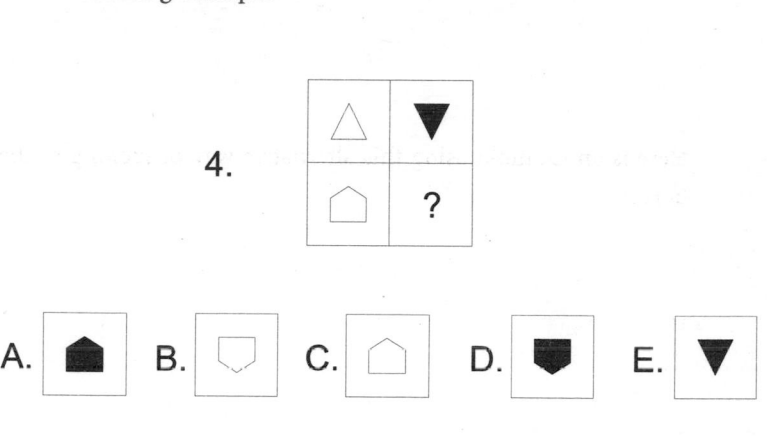

In this example, the first box in the top row has a white triangle. The second box in the top row has a *reversed black* triangle. So the rule is that you reverse the shape and you change its shade. If you look at the first box in the bottom row, it has a white pentagon. If you follow the same rule as for the first row, you are looking for a reversed black pentagon, so the correct answer is option D.

3×3 table

Another way of presenting the data is in a 3×3 table. There is usually a theme in each row, or in each column, with one box missing. This is the box that you are asked to complete by choosing from a number of boxes shown beneath the table. An example of a 3×3 table item is shown next.

5.

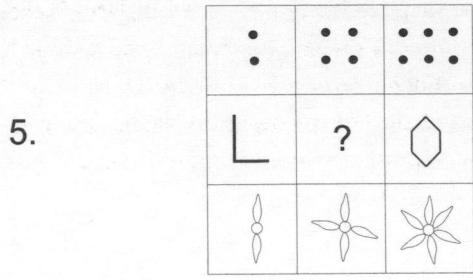

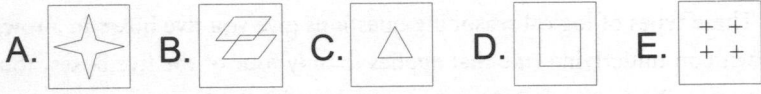

The top row contains dots. The first box contains two dots, the second four and the third six, indicating that each box has two more dots than the previous one. The third row, which is also complete, contains images of a flower. The first box has two petals, the second four and the third six petals, following the same rule of adding two to each box. Since the 'add two' rule applies to both the top and bottom rows, it must also apply to the middle one. Look at the middle row and you will see that it contains shapes. You are therefore looking for 'something' of which there are two in the first box and six in the last box. You can see that the shape in the first box has two sides and the last shape has six sides. So the 'something' that is increasing by two is the number of sides. You are thus looking for a shape with four sides. The only one of the five possible answers that has four sides is D.

This example contains a distracter in one of the five possible answers. Distracters are things that might confuse you when trying to figure out the correct answer, and they will be discussed in detail later in this chapter (see page 148). Box E from the five possible answers contains four *shapes*, which might confuse you, as you know you're looking for four of 'something'. You must remember the rule: you are looking for four *sides*, not four shapes, and that's because the boxes in the middle row contain one shape each and the 'something' that is changing is the number of sides of each shape. Therefore, if you choose box E, your answer will be considered wrong.

A variation of the 2×2 or 3×3 questions are items that consist of 16 boxes (4×4). These are usually more complex and are used less frequently in speeded recruitment tests.

Which figure is the odd one out?

These types of logical reasoning questions give you five boxes in a row, with an underlying rule that applies to only four of the five boxes. You need to find out what the rule is and identify the figure that does not follow the rule. This will be the odd one out. Can you spot the odd one out in the example given below?

6.

If you look at the number of items in each box, you will see that boxes A, C, D and E each contain three shapes. It is only box B that contains two shapes. Box B is therefore the odd one out.

When identifying 'the odd one out', you are looking for all sorts of things. It might be that one box has shaded shapes whereas the shapes in the other boxes aren't shaded. It might be a box that contains shapes with corners, while the rest of the boxes contain rounded shapes. You might be given a box with one large shape inside, while the other boxes have small objects in them. Or you might be looking at the number of shapes, corners or sides.

Which figure belongs in neither group?

These are similar to the 'Which figure is the odd one out?' questions explained above. In this type of question, you are given groups of items and you are looking for the figure which does not belong in either group. You are typically presented with two groups, each of four boxes containing shapes. The shapes are grouped together because they have something in common – an underlying rule.

Once you work out what the rule is, you will be able to identify, from the four response options given beneath the question, which one contains shapes that do not belong to either of the two groups. This may sound confusing, but the example below will help you to understand it. Try to find what the common rule is between the shapes in Group 1 and Group 2 respectively.

7.

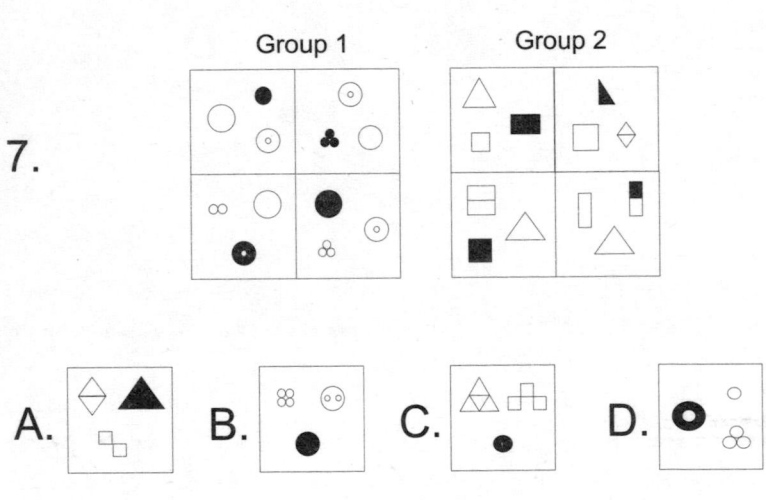

You will observe that Group 1 contains 'only circles' and Group 2 contains 'no circles but shapes with corners'. These are what stand out and these are your rules. Now that you have found the rules, look at the four possible options below. You can see that box A contains only shapes with corners and could therefore belong in Group 2. Boxes B and D contain only circles, and could therefore belong in Group 1. It is only answer C that has both circles and shapes with corners. Box C is therefore the odd one out – the one that doesn't belong to either group and the answer you are looking for.

Which figure completes the grid?

In order to understand this type of question, picture the data presented to you as a puzzle with one part missing. When completed, the puzzle will give you a logical pattern of sixteen boxes. Once again, the possible missing pieces are given under the uncompleted grid.

Try to find the logic of this puzzle from the pieces given to you. Can you observe any patterns in the first two rows?

8.

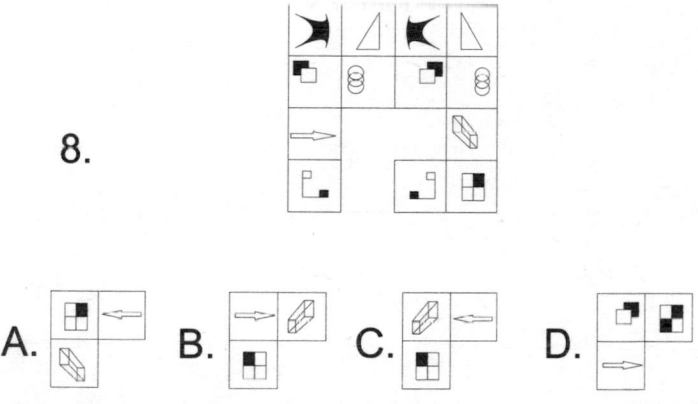

You might have noticed that there are some similar shapes in the first two rows. For example, looking at the top row, the first and third boxes contain similar shapes, as do the second and fourth boxes. The same happens in the second row. Now you need to work out how the shapes are similar. It's easy to see that the images are mirroring each other. The first box of the first row is the mirror image of the third box in the first row. If this applies to the rest of the boxes, you have your rule. Try it out. Indeed, the second box in the first row is the mirror image of the fourth box in the first row. The same applies to the second row.

In order to find the correct answer from the four possible options given, you must apply the same rule. The first box of the third row must be the mirror image of the top right box you are looking for. You are therefore looking for an arrow, facing left. You can see such an arrow in both options A and C, so you need another clue. The box missing from the fourth row must be the mirror image of the fourth box in the fourth row. Of your two possible answers, only C contains the image you are looking for in the lower box. Now that you have your answer, check the remaining box, just to make sure you've come up with the correct rule. The upper left box in answer C is the mirror image of the last box in the third row, so answer C is definitely correct.

The basic rules

The types of logical reasoning questions described above are the most common ones you will come across if asked to sit a test. The vast majority of logical tests use the 'Which figure completes the grid?' type of question. If you feel comfortable with the different forms of presentation of the previous examples, it's a good idea to get acquainted with the basic rules.

We have divided the basic rules into three different parts to make them easier to grasp. First we will look at the **mathematical ability** that is required to sit a logical reasoning test, and in what way mathematics can be used in such a test. We will then explain how **shapes** play a very important role in the presentation of the questions. And then we'll look at how **shades** are involved in these tests. Finally, we will tell you the secrets of **distracters** and how you can avoid falling into their traps.

Mathematics

How high is your mathematical ability? Not too good? No problem. You only need to know the absolute basics of mathematics to score well in a logical reasoning test. You need to know a little bit of addition, subtraction, multiplication and division and how to find square roots. Having read the chapter on numerical tests, this will seem like a piece of cake. In this section we have also included the 'sequence' form of questions.

Addition

When it comes to adding, you only need to add one 'something' at a time (in most cases). But you need to identify what this 'something' is. For example, the simplest form would be adding the number of shapes. A slightly more complicated form would be to add sides to a shape, or to add corners. An even more complicated form would be to add the number of letters within words. Take another look at item 1 (page 112), for example. The rule was to add one shape in each box. In item 2, the rule was to add one shape around the other or others. In item 3, it was to add one side each time. Let's see if you can find what it is you are asked to add in the following question.

9.

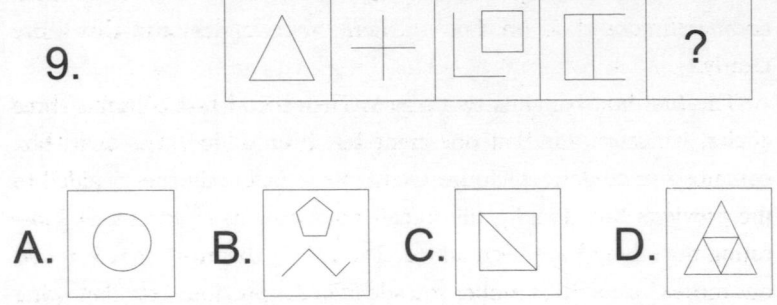

A. B. C. D.

In this example you need to add corners; there is one more corner in each consecutive box. Starting with the first box, you count three corners in the triangle. Moving on to the next box, there are four corners in the cross. The third box contains two shapes: the square with four corners and the letter L which has one corner. Added together, these two shapes give you five corners. Box 4 has a shape with six corners, four inside the rectangle and two formed by the protruding lines. The fifth box must therefore contain one or more shapes with seven corners altogether. Looking at the four possible answers, A has no corners whatsoever, B has seven corners, C has six corners and D has twelve corners. The correct answer is B.

10.

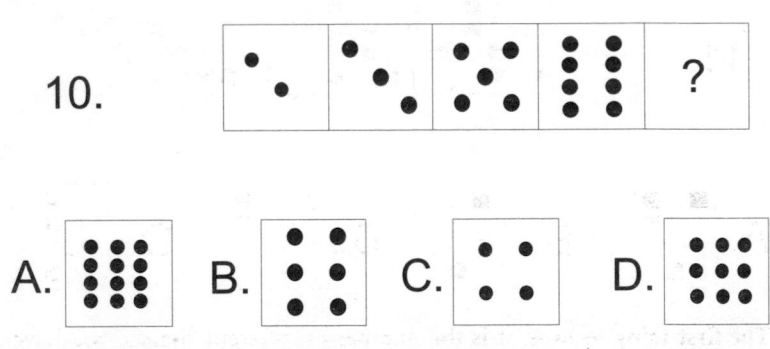

A. B. C. D.

A more difficult form of addition would be to add more than one of 'something'. So instead of consecutively adding one item (be it a shape, shade or corner etc.), you may be adding one, then adding two, then adding three and so on. Look at item 10 to understand this more clearly.

The first box contains two circles. The second box contains three circles, which means that one circle has been added. The third box contains five circles, which means that two circles have been added to the previous box. Finally, the fourth box contains eight circles, indicating that three have been added. The rule is therefore that you add one more circle to the number you added in the previous box. Following this rule, you know that you need to add four circles in order to find the next box in the sequence. Eight plus four equals twelve, which is only found in answer box A.

Subtraction

Subtracting shapes, corners etc. follows the same rules as adding 'something'. The only difference is that the 'something' is being removed from each box. Try the following example. It is quite complicated and it would be considered a difficult question if you didn't know which method to use. But in this instance you know that you are subtracting something; try to find what you need to subtract.

11.

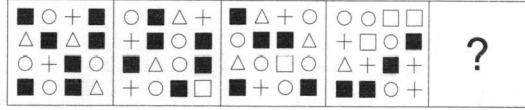

A. 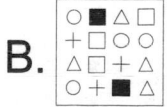 B. 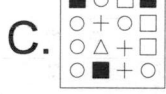 C. 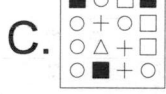 D.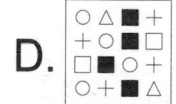

The first thing to look at is the number of different shapes. You have a circle, a cross, a black square, a white square and a triangle. Now look at how many shapes there are in each box.

	BOX 1	BOX 2	BOX 3	BOX 4
CIRCLE	4	4	5	4
CROSS	2	3	2	4
BLACK SQUARE	7	6	5	4
WHITE SQUARE	0	1	1	3
TRIANGLE	3	2	3	1

The only shape that is decreasing in number as the boxes proceed is the black square. Since the last box contains four black squares and you have seen that it decreases by one in each box, the answer must contain three black squares. The correct answer is therefore C, since it is the only one with three black squares.

Multiplication, division, square roots

Multiplying, dividing and finding the square root are the highest levels of maths you will be asked to perform in a logical reasoning test. You might be shown a number of 'something', such as dots or shapes, all of which are multiplied (e.g. doubled) or divided by a number (e.g. halved), or they may be numbers to a power, or square roots of numbers.

Try to work out the rule behind the following example.

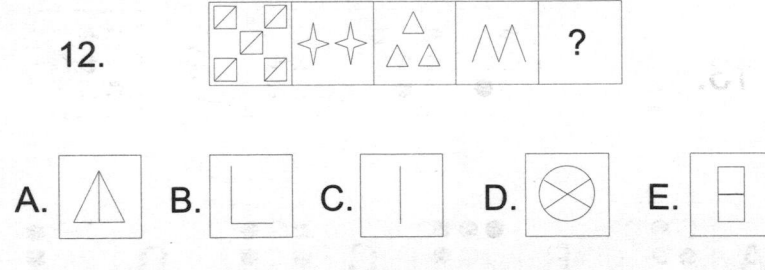

12.

Since the four boxes contain simple forms of shapes (with no shades or colour etc.), you can look at the number of shapes, the number of sides or the number of corners. Let's see how many there are of each in each box.

	BOX 1	BOX 2	BOX 3	BOX 4
NUMBER OF SHAPES	5	2	3	1
NUMBER OF SIDES	25	16	9	4
NUMBER OF CORNERS	30	8	9	3

The most striking group is the number of sides. The numbers are 25, 16, 9 and 4, all of which are numbers to the second power, i.e. 5^2, 4^2, 3^2 and 2^2. The number that logically follows is 1^2, which equals 1. So the correct answer is C.

Sequences

Finally, you might be asked to identify the correct box to complete a series of boxes that follow a specific sequence. In a five-in-a-row item, a sequence is formed by two numbers that interchange between them, i.e. 1-2-1-2-1. In a 3×3 table the sequence could be 2-4-6 in each row, giving the sequence of 2-4-6-2-4-6-2-4-6 in the whole table. Can you find the sequence in the following example?

13.

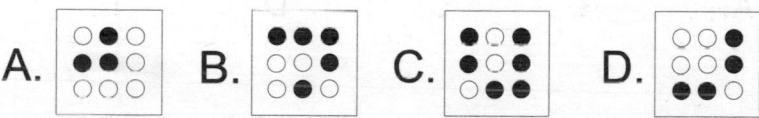

In this example, there are nine circles in each box. In the first box three of the circles are black, in the second box five, in the third three and in the fifth three. You are asked to identify the fourth box which, according to this rule, must contain five black circles. The correct answer is found in box B.

That's as far as your mathematical ability needs to go, in order to solve logical reasoning tests. The next thing we will see is how shapes can be used in logical items.

Shapes

Shapes can do all sorts of things inside a box. They can be added or subtracted, they can move, they can be mirrored, or they can have other shapes put above, beneath, inside or outside them. They can be split into two or more shapes; they can be split and moved around in the box. Similarly they can be merged, or merged and moved around in the box. They can be altered in shape or they can be placed symmetrically within the boxes. We will look at all of these separately and take you through the logic behind them.

Moving in space

In these situations, each of the boxes contains shapes that move with a specific order within the space of the box. All you need to do is trace the path of each shape or object and work out what the next move will be. The greater the number of shapes, the harder the question, because you need to figure out the next step for each shape. Here is an example with three objects; can you work out the answer?

14.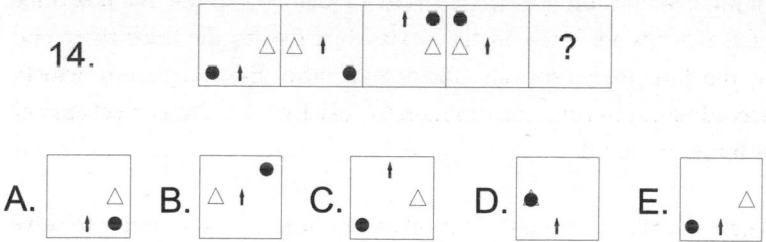

A. B. C. D. E.

There are three items in each box. As the sequence progresses, you will notice that these items move within the space of the box. Look at how each one moves around. The circle moves anticlockwise, covering with each move the space to the next corner. The triangle moves horizontally, from one side of the box to the other. The third object, the arrow, moves vertically, covering with each move half the space to the other end of the box. If you trace the next move of each shape, you expect to find the circle in the lower left corner, the triangle in the right middle of the box and the arrow in its initial position, the lower middle of the box. The correct answer is option E.

Rotating

Some logical reasoning items involve rotating the shapes presented. You will need to spot the images rotating clockwise or anticlockwise, following, most typically, rotations of 45, 90 or 180 degrees. Occasionally rotations may be a slight turn with no definite degree rule.

Look at the rotation example on the following page.

15.

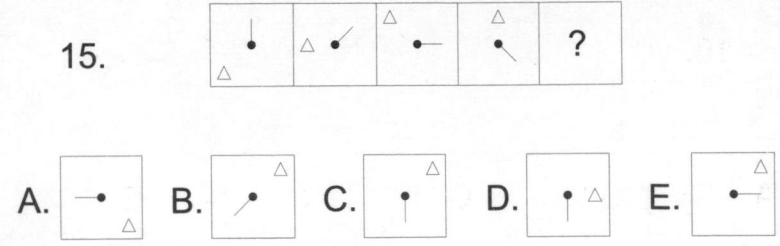

A. B. C. D. E.

Here you can see one shape moving in space and one shape rotating. The triangle is moving clockwise, each time covering half the space to the next corner of the box. The line is rotating clockwise, around the dot in the middle of the box. In the first box the line is pointing directly upwards, moving 45 degrees in each consecutive box. Following these two rules, the correct answer is option C, where the triangle is located in the upper right hand side of the box and the line is pointing directly downwards.

A more complex form of rotation involves a number of images within each box, each rotating independently. For example, one image could rotate 90 degrees clockwise, a second one 180 degrees and a third one 45 degrees anticlockwise. The correct answer would be a box which has each image correctly rotated, according to the previous boxes.

Look at the following example; can you spot the correct answer?

16.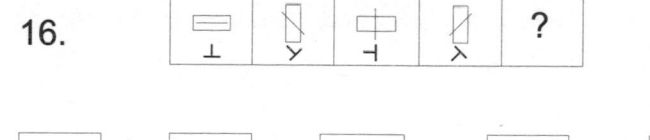

A. 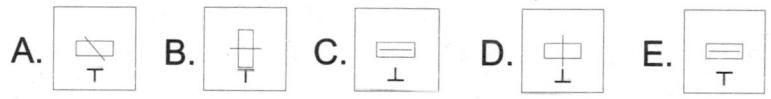 B. C. D. E.

Here you have three shapes. The line rotates 45 degrees clockwise, the rectangle rotates 90 degrees (clockwise or anticlockwise – in this case it is the same) and the T shape rotates 45 degrees anticlockwise. Following the moves of each of these shapes, the correct answer is option E.

Mirroring

When it comes to mirroring, you need to be able to look at the contents of the boxes as if you were looking at them through a mirror. In the following example, the upper right box is the mirror image of the upper left box, which gives you the rule to solve it: the lower right box should be the mirror image of the lower left box.

17.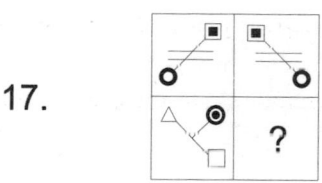

A. 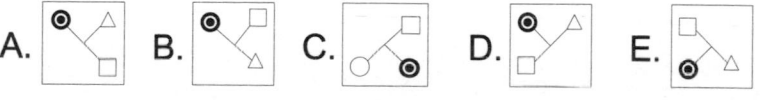 B. C. D. E.

It is only option D that is the exact mirror image of the lower left box.

Another way an image can be mirrored in a 2×2 table is vertically. This means that the lower left box would be the mirror image of the upper left box, so that the blank lower right box must be the mirror image of the upper right box.

A more complicated way of mirroring images is in 4×4 tables (see item 8 on page 121 as an example), in which the mirrored images are not located in adjacent boxes. Some possible locations are the following:

Row 1:	1	2	3	4
Row 2:	5	6	7	8
Row 3:	9	10	11	12
Row 4:	13	14	15	16

- Row 1 could be the mirror image of row 4 (i.e. box 1 mirrors box 13, 2 mirrors 14, etc.) and row 2 of row 3 (i.e. box 5 mirrors box 9, 6 mirrors 10, etc.).
- Row 1 could be the mirror image of row 3 (i.e. box 1 mirrors box 9, 2 mirrors 10, etc.) and row 2 of row 4 (box 5 mirrors box 13, 6 mirrors 14, etc.).
- A more complicated form of mirroring would be if the corrosponding images were located diagonally (i.e. box 1 mirrors box 6, 2 mirrors 5, 3 mirrors 8, 4 mirrors 7, etc.).

All of the above methods of mirroring can be applied to columns as well.

Overlapping

The rule of overlapping involves shapes being added to boxes, each one shown either on top of, or beneath, the other. The box you are looking for must contain all the shapes shown in the previous boxes, in the right order, plus one shape added. There is usually a rule involved on how the shapes are added, i.e. 'each shape is placed on top of the other' or 'one shape is placed on top, the next one beneath, then on top and so on'. Look at the example given below and try to find the correct answer before reading its explanation.

18.

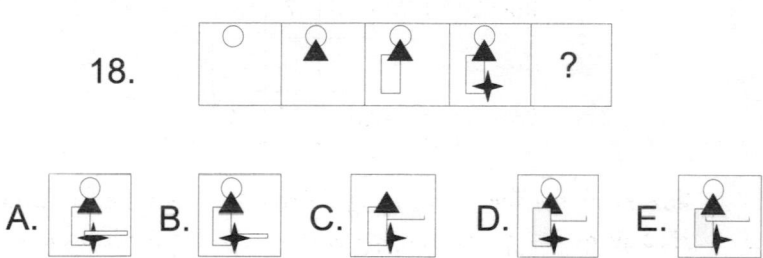

The correct order must be one circle, then a triangle on top, followed by a rectangle beneath the triangle and then a star on top of the rectangle. It is of no importance which shape will come next, as long as it is beneath the star. The correct answer is box B.

Eliminating

This rule involves eliminating *exactly the same* sides or shapes that fall onto one another. Look at the next example. You are given a plain square in one box and a square with an X in the second box. These two shapes put together will result in the elimination of the square (since it is found in both boxes), leaving just the X.

You have to pay attention to the 'exactly the same sides/shapes' part of the rule of elimination. The sides or shapes that are eliminated must be located in exactly the same position in both boxes. If they are not, then they are not eliminated but they are both found in the third box. Look at the two examples below.

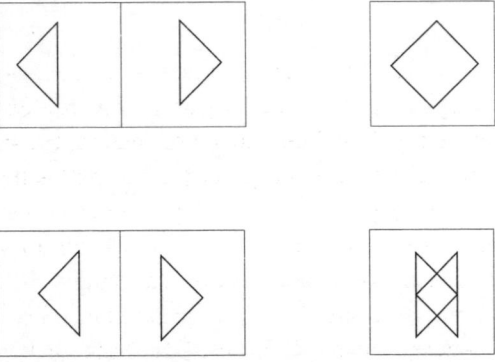

In the top example, the two vertical sides of the triangles are located exactly in the middle of the box, therefore eliminating one another and leaving a rhombus in the third box. In the second example the whole triangle shape is located in the middle of the box; the vertical sides are therefore not in the middle – one is slightly to the right and one is slightly to the left. Therefore they do not eliminate one another, instead resulting in the shape shown in the box on the right.

An example of an elimination item is illustrated below.

19.

A.   B.   C.   D.   E.

You can see that the third column is the outcome of elimination if columns one and two are put together. Let's look at the first row. In the first box there is a square and a circle. The second box contains two of the square's sides and the same circle (same size, same location within the box). If you eliminate the common parts from both boxes (in this case the two lines and the circle), what remains is illustrated in the third box. The same applies to the second row. In this row, the square and the cross do not eliminate each other because they have no common sides, resulting in a cross-within-a-square, in the third box. The black circle is eliminated because it is found in both boxes one and two of the second row, in exactly the same position. Finally, as far as the triangles are concerned, putting them together will give us a rhombus, and the vertical side of both of them has been eliminated because it is in the same position in both triangles. The vertical line in the first box and the horizontal line in the second box do not eliminate each other, giving us a big cross in the middle of the third box. Following the same steps for the third row, the answer you are looking for is option A.

Splitting

Splitting requires the ability to visualise an image in its component parts. How many components can you find in a simple square?

Square: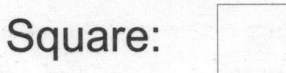

A. B. C. D.

A. You can have four separate lines, two vertical and two horizontal.
B. You can have one separate line and three connected lines.
C. You can have two 'L's, one inverted.
D. You can have two separate lines and two connected lines.

If you can find so many ways of splitting one simple square, imagine how many components you can find in more complex shapes!

Below you are given an example which involves simple splitting. Try to see what happens to the shape in the first row in order to figure out what will happen to the shape in the second row.

20.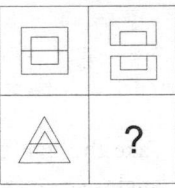

A. B. C. D.

In this example, the top left box contains a shape made out of two squares, one inside the other with a horizontal line across their middle. In the box to its right, the shape is cut in half, exactly where the horizontal line was.

The lower left box contains two triangles, one inside the other with a similar horizontal line in the middle. Following the same rule, the shape must be divided into two shapes, cut exactly where the horizontal line lies. The correct answer is option C.

Merging

Merging is the exact opposite of splitting. In merging, you are presented with shapes which are then combined. If you have a 3×3 table as shown in the example below, the first two columns 1 and 2 of each row are merged to make the third. Before you read the answer below, try to work it out yourself.

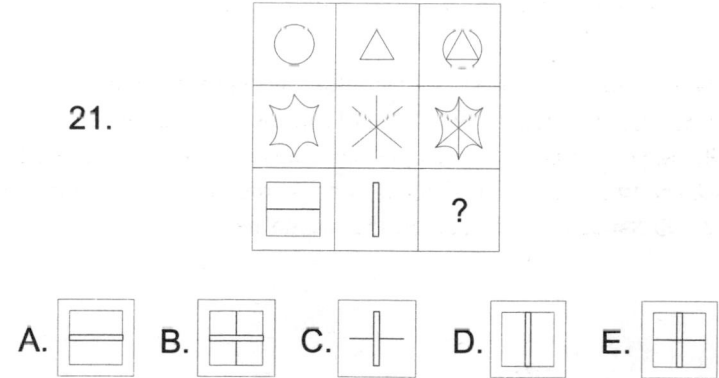

21.

Look at the first row. Take the first box, the circle, and merge it with the second box, the triangle. The outcome of this merging is shown in the third box of this row, which is a triangle within a circle. Now look at the second row. If you merge the first two boxes, you get the shape shown in the third box. If you apply this same rule to the third row, you will find the answer you are looking for is option E.

Altering shapes

This form of question involves a slight alteration in the shapes presented. Altering a shape could mean all sorts of things. It might be pulling one side of it to the left, to the right, upwards or downwards. It might be changing one side from being round to having a corner or a pointing edge. It could be making the shape bigger or rounder etc. You need to find what this alteration is, so that you can apply it to the shape in question.

Look at the example below. You are given a 2×2 table, with a square in the first box of the first row. In the box to its right, the shape has been altered by a little black circle protruding exactly in the middle of its right-hand side. Apply this rule to the oval shape given to you in the second row. Visualise what the shape would look like, if altered by 'a little black circle protruding exactly in the middle of its right-hand side'.

22.

A. B. C. D.

The correct answer is option B.

Symmetry

In a symmetry question, you have to find the symmetry in the boxes given to you, by looking at them as if they all formed just one picture altogether. So don't look at each box separately. The most important thing to do is to find the line that divides the total 'picture' into two parts. If you have a 3×3 table, this division can be the middle row or the middle column. If you have five boxes in a row, the division might be the middle box. If you have four boxes, the symmetry line might be between the second and third boxes.

Have a go at the following symmetry question:

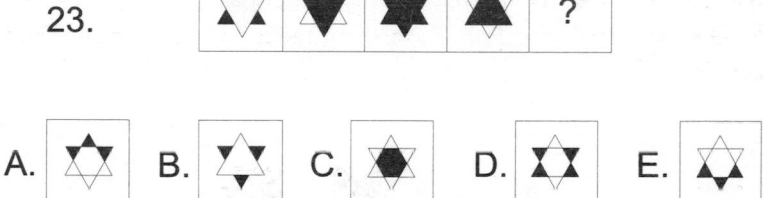

23.

In this example the symmetry lies in the middle box in which the star, or the two triangles (one pointing upwards and one pointing downwards), is coloured black. Look at the two boxes on either side of the middle one. What do you observe? In the left box, the black triangle is in front of the white one and is pointing downwards, whereas in the right box, the black triangle, while still in front of the white one, is pointing upwards. Now look at the remaining box on the far left. The white triangle is in front of the black one and is pointing downwards. The rule of symmetry allows us to guess that the empty box must contain a white triangle, in front of a black one, pointing upwards. That is option B.

Shades

Now that we have seen what shapes can do, let's look at how shades can be used in a logical reasoning item.

Adding shades

If you have understood the 'addition' rule as far as shapes are concerned (see page 123) it will be very easy to figure out how the 'adding shades' rule applies. You might be presented with a question like the one shown below. Here you can see that in the first box nothing is shaded. The second box has the outer circle shaded; the third has one more circle shaded. You are asked to find the fourth box, but you are also given the fifth, in which everything is shaded. The rule is that each box has one extra part shaded (an inner circle, working inwards, in this example).

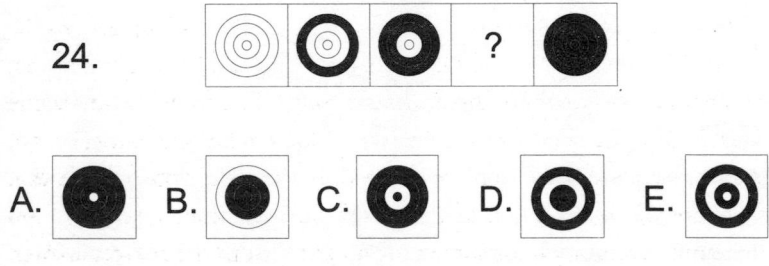

The correct answer is A, which has everything but the small middle circle shaded.

Questions that follow this rule can therefore have shades added to one part, or, in more complex items, more than one part, from one box to the next.

Removing shades

Removing shades follows exactly the opposite rule to adding them. When the item contains five boxes in a row, the first box usually contains a fully or partly shaded shape. As the boxes proceed, you notice that each time at least one shaded shape is removed. This means that the correct answer is the one containing the shape of the last box presented to you (usually the fourth box) with at least one shaded shape removed. Look at the following example:

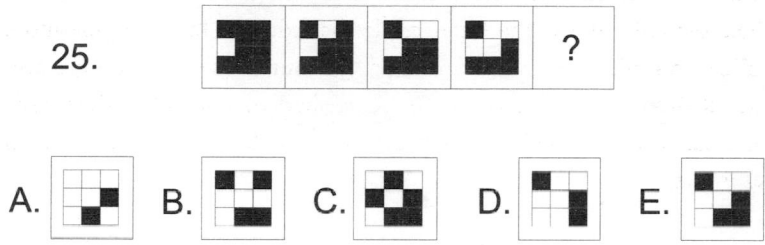

In the first box you have one white (not shaded) small square. In the second box you have two, which means that one shade has been removed; in the third box you have three; and the fourth box contains four white squares. You are therefore looking for a shape with five white, or not shaded, squares. If you look at the possible answers you will see that there are two options – B and E – with four not shaded squares. You therefore have to look for something more to get to the correct answer.

If you look closely, you will see that the squares that turn white are not selected at random each time, but they remain the same; in every box, one of the *remaining* black squares turns white. Look closely at answer B. The top right square is shaded. However, this square already turned white in the third box. It cannot be shaded again. This leaves you with the correct answer: option E.

Note that apart from the rule that 'one of the *remaining* black squares will turn white' there was no specific rule about which of the black squares would be 'un-shaded' next. This may have caused you to spend time looking for a rule that wasn't there (i.e. looking for a pattern in which square becomes un-shaded from the remaining black ones). But as you can see from the answer, you didn't need to figure out such a pattern, because the rules that you had already identified only left you with one correct answer.

TIP: When you have difficulty identifying all the underlying rules in an item, you should bear in mind that you may already have all the rules that you need. It is worth checking whether the rules that you have already identified exclude all the response options apart from one. If so, you have your answer.

Moving shades

In this rule, it is the shades that move either within a box or within a shape. Look at the example below:

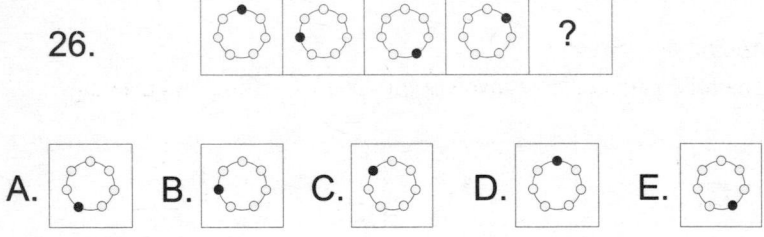

In this example you have seven circles attached to a string. In the first box, the highest circle is the one with the shade. In the second box, the shade has moved two circles anticlockwise. Looking at the next boxes, the shade always moves two circles anticlockwise. So the correct answer is option C.

Combining rules

So far, you have been shown the basic rules in logical reasoning items. We have explained the logic behind each one and given you examples of how to put the theory into practice. This is, however, only the first step. Although items are often based on just one rule, psychometricians usually combine more than one rule within an item, to make it more complex. So the next step is to learn to identify the basic rules when combined in a question. For example, we have looked at how shapes are added and also how shapes are rotated. Could you identify these two rules if you were to see them combined? This section explains how the basic rules can be combined and gives you examples to practise on.

The secret to solving items with combined rules is to follow each rule separately. Then combine what you would expect to see if you were to follow each rule separately and that will give you the correct answer. This is not as difficult as it sounds. Look at the examples below and you will see.

Adding and moving
The following example involves the rules of adding and moving.

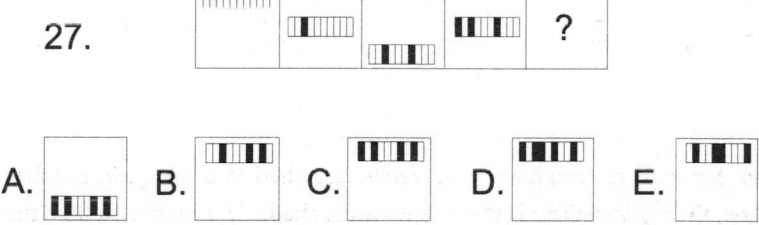

As the boxes progress, one shade is being added each time. There are two things to note here. Firstly, if a shade is added, it cannot be removed. And secondly, there is no rule as to which small rectangle will be shaded next. There could have been a rule that the adjacent rectangle is shaded each time, or one is shaded and the next is left un-shaded. However, there is no clue in this example to show you which rectangle will be shaded next.

You might also notice that the shape containing the shades moves inside the box. It moves vertically, covering half the space each time, first downwards and then upwards. To find the box that completes the sequence, try to look at each rule separately.

In the first box, there are no shades whatsoever. The second box has one shaded rectangle, the third one in the row. The third box has the rectangle which had already been shaded, plus one extra, the seventh in the row. The fourth box has the previous two rectangles shaded plus the first rectangle. The correct answer must therefore have all the previous rectangles shaded, plus one more. Two of the answers follow this rule, options A and C.

Now it is time to look at the second rule, moving in space. From the two possible options A and C, it is option C that has the shape in the expected position, that is, at the top of the box.

Subtracting and rotating

In this type of combination question, shapes are removed from each box and the remaining ones are rotated. Again, try to follow each rule separately. The box in question must contain one object or shape fewer than the last box presented to you (usually the fourth box). It must also follow the rotation that logically continues from this box.

28.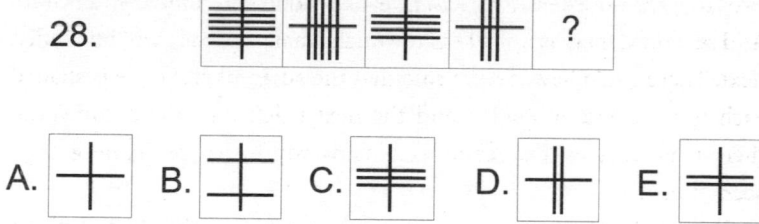

In the first box, there are six horizontal lines and one vertical. In each of the following boxes, one of the six lines is removed, but the single line is always present. Furthermore, the whole shape is rotating by 90 degrees in each box. You therefore have five vertical lines and one horizontal line in the second box. In the third box you are presented with four horizontal lines and one vertical line. Finally, in the last box given to you, you see three vertical lines and one horizontal line. This means that the correct answer must have three lines, two horizontal and one vertical. The correct answer is option E.

Splitting and moving or rotating

When an item involves splitting and moving, you expect the shapes within the boxes to be split and then moved or rotated. Look at the example on the next page. Here's a tip to help you work out the rule: the first box in the first row does not contain two shapes, a patterned circle and an arrow. It actually contains three shapes: a patterned circle, a line and an arrowhead!

29.

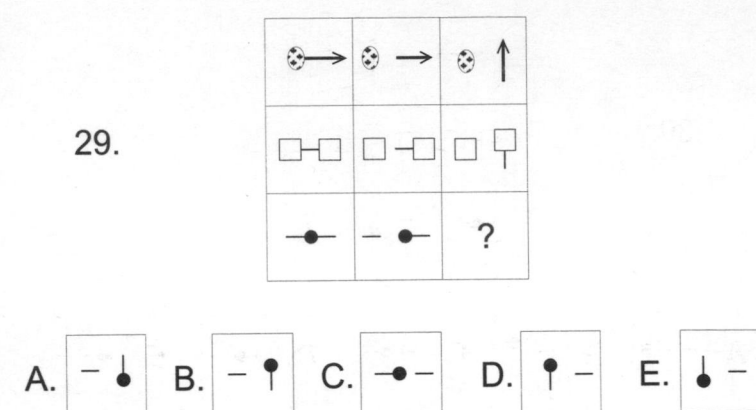

A. E. and others...

In the first box you have a patterned circle, a horizontal line and an arrowhead, all attached together. In the next box the circle separates from the other two shapes, the line and the arrowhead, which remain attached. In the third box the line and the arrowhead, still attached together, are turned anticlockwise.

The second row follows exactly the same rule. The three shapes that appear in the first box are two squares and one horizontal line. The second box, as expected, separates the first square from the line and the second square, which remain attached together. In the third box the line and the attached square are turned anticlockwise.

So here comes the one you are asked to solve. Can you see three items in the first box of the third row? Two horizontal lines and one circle. In the second box, the first line is separated from the circle and the second line, which remain attached. In the third box the attached items are always turned anticlockwise. Looking at the possible answers underneath, you will find the correct option is box A.

Next is a similar example.

30.

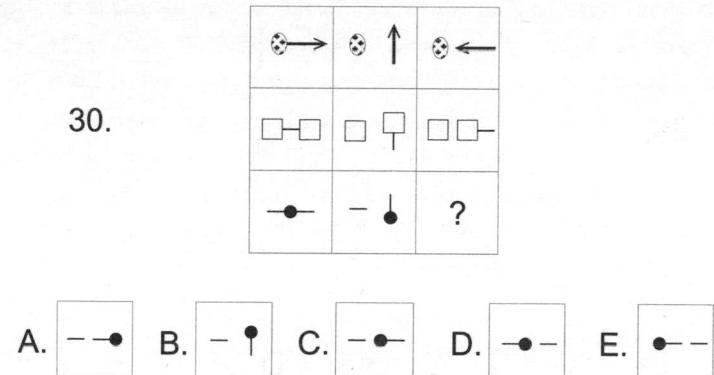

Like in item 29, the rule in this example is that the shape is first split, leaving the last two parts combined and turned 90 degrees anticlockwise (for the second box). Then the shape that has been rotated, is rotated again 90 degrees anticlockwise. The correct answer is therefore option A.

Merging and moving

Merging and moving involves merging the components of the boxes and also moving them within the box. Look at the example below and try to work out the solution before reading it.

31.

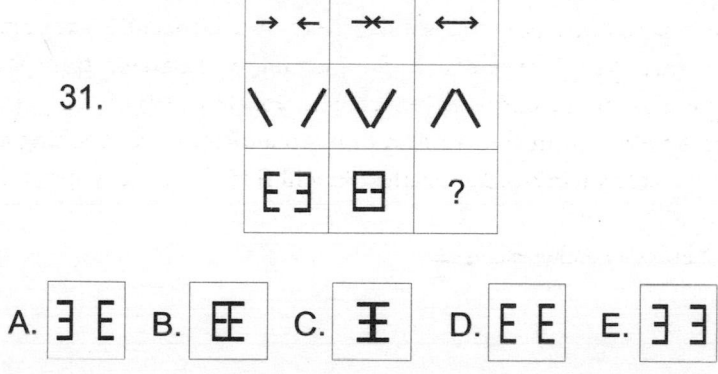

If you look at the first box of the first row, there are two arrows facing one another. In the second box the arrows have merged by slightly touching each other with their pointing ends. In the third box they have moved past one another, but are still touching with their back ends. The same rule applies to the second row. Therefore, the rule is that the two shapes are merged by slightly touching and then continue in the same direction moving away from one another, but are still touching with their back ends. The correct answer is option C.

By now you have a pretty good idea of what types of combinations you may be faced with when asked to sit a logical reasoning test. Essentially, you can combine any rules and sometimes you can combine more than two. You will find some more examples in the practice questions in the next chapter.

Distracters

If you think you have understood all of the above rules and combinations, you are ready for the next step in mastering your opponent: learning about distracters. The distracters do exactly what their name suggests they do: they distract you. They are meaningless shapes, figures, images or shades that are only there to confuse you. They are tricks that stall you from finding the correct answer. Psychometricians use distracters to increase the level of difficulty in logical reasoning tests. Some distracters are simple and easy to spot; others are more complex.

You might be reassured to know that facing difficult distracters in a speeded test is the exception rather than the rule. Since you will have a limited time in which you can solve logical items, you will mostly be assessed in identifying and applying rules rather than in realising that you are looking for a rule that does not exist.

You can have shapes as distracters or shades as distracters, or you might have a sequence starting from a 'less logical' number. Let's look at each one separately.

Shapes as distracters

The best example to look at for shapes as distracters is item 11, which you saw earlier in this chapter:

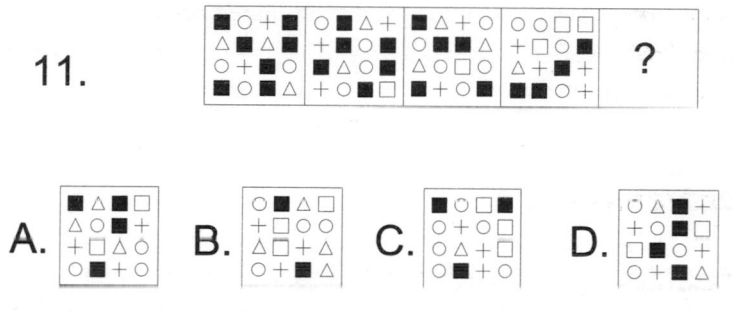

You are given lots of shapes but you only need to focus on the shaded squares. The answer is pretty obvious when you know that the shapes to look out for are shaded. What if the shapes you are looking for do not immediately stand out from the rest? Would you be able to find them among the distracters?

Try to solve the example on the following page.

32.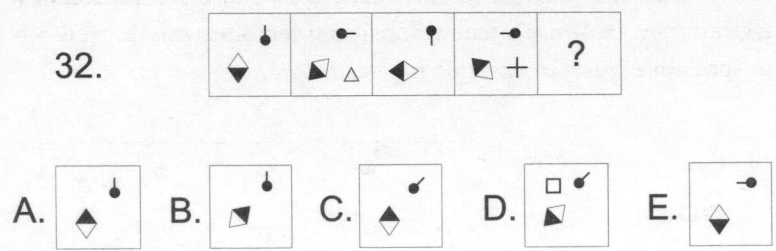

In this item you can observe that the circle with the line attached to it, found in the upper right-hand corner, is rotating clockwise. The half-shaded rhombus in the lower left-hand corner is also rotating clockwise.

While you are focusing on two rotating figures, you suddenly see another shape appearing out of nowhere! In the second box it is a triangle and in the fourth box it is a cross. Try to see if these shapes have any meaning – if they appear in the subsequent boxes following a specific rule. If not, then they are clear examples of distracters. In such a case, pay no attention to them and simply follow the rotation rule that you have spotted. You have to check with the answers though!

Following the rotations of the circle with the line and the rhombus, the next logical step is only found in option A. You therefore know that the other shapes were just distracters. If there are two correct answers based on the rules that you have identified, then you should check to find a rule for the extra shapes, but if you are left with only one possible response, you know you have solved the problem.

Shades as distracters

If you have worked out that the rule of an item is 'moving in space' and you see that some figures have been shaded, but no shade-rule seems to apply to these figures, then there may be shade-distracters involved. But you have to be sure that the shading does not follow a logical sequence. You can then ignore it and continue with the moving-in-space rule. Such an example is given below:

33.

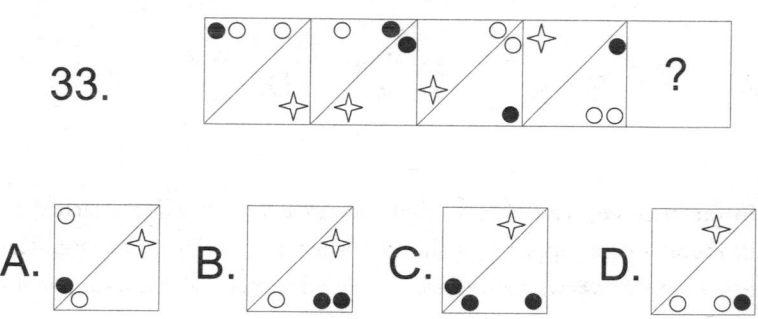

In this example you can find the solution, which is option D, regardless of the shaded circles. There is no rule behind the shades (there is one shaded circle in the first box, two in the second, one in the third and one in the fourth). Furthermore, if you follow the moves of each item in the boxes, D is the only solution possible. So the shades in this example were only there to confuse you and to make the answer more difficult to find.

Distracters in sequences

In order to have a sequence, you usually add one of 'something' and subtract one of 'something' (be it a shape, shade or corner etc.), giving you a sequence of, for example, 2-3-2-3. Usually the sequences start from a small number, such as 1-2-1-2 or 2-4-2-4. It is possible, however, for the sequence to begin with a less 'logical' number, like 6, and follow the pattern of 6-4-6-4.

In the following example you are presented with words, the meanings of which you won't understand unless you are familiar with the

Greek alphabet and language. Since this is not a verbal test, don't waste your time looking for meanings, opposites, nouns or verbs (this is why the words are presented in a form you might not be familiar with.

34.

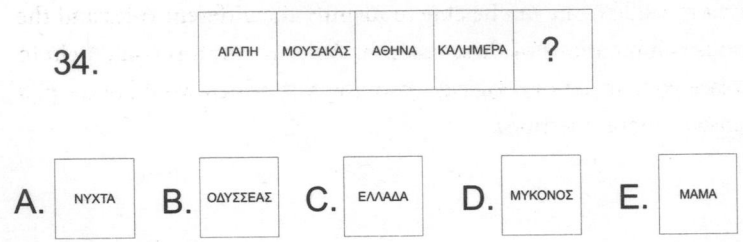

ΑΓΑΠΗ	ΜΟΥΣΑΚΑΣ	ΑΘΗΝΑ	ΚΑΛΗΜΕΡΑ	**?**

A. ΝΥΧΤΑ   **B.** ΟΔΥΣΣΕΑΣ   **C.** ΕΛΛΑΔΑ   **D.** ΜΥΚΟΝΟΣ   **E.** ΜΑΜΑ

What do words consist of? Letters. What can you do with letters? You can count them! The first box contains the word 'ΑΓΑΠΗ', consisting of five letters. The word 'ΜΟΥΣΑΚΑΣ' in the second box has eight letters, 'ΑΘΗΝΑ' has five and finally 'ΚΑΔΗΜΕΡΑ' has eight letters. So you have your sequence: 5-8-5-8. The next in the sequence is 5, so you are looking for a word with five letters. From the possible answers, only the word 'ΝΥΧΤΑ' has five letters, which is the correct answer.

Conclusion

We have seen in this chapter some of the most common rules found in logical reasoning questions. You have been shown how high the level of your mathematical ability should be in order to solve these questions, in what way the shapes can be altered or moved within the boxes, what questions you might be faced with involving shades, and how symmetry and sequences can be used as rules. We have also looked at the ways that different rules can be combined to increase the complexity of the questions. And you are now aware of distracters that may be there to confuse you and which need to be ignored.

In the next chapter, you will find a number of practice questions. These vary in complexity but essentially they begin with simple ones and become more difficult as you proceed. Practise these questions and check your answers at the end. Remember that the more you practise, the more readily you will be able to identify the different rules and the various combinations of these rules. In the long run this could help to increase your speed in a logical reasoning test, which would allow you to answer more questions.

Step 5: **Face your opponent**

You have now gone through the steps of meeting, getting ready for and mastering your opponent. The next step is to face your opponent.

This chapter includes a numerical reasoning test and a logical reasoning test. Your work so far should mean that you're feeling prepared for the tests, so try to complete them under test conditions. This means you should take each test in one sitting, trying to minimise any interruptions, concentrating hard and giving each one your best shot.

Even though there is no time limit given for these practice tests, try to complete them as quickly as you can, as speed will be important when you take the real tests.

Practice numerical test

Instructions

This is a test assessing your ability to use information presented in tables and charts to answer numerical reasoning questions. Each table or chart is followed by eight multiple-choice questions. For each question you are given five response options, of which only one is correct. Your task is to use the information given to derive the correct answer.

You may use a calculator and rough paper in this test.

FIGURE 1: **Working population**

	2005		2007	
	Men	Women	Men	Women
Self-employed workers	300	200	340	170
Other-employed workers	3,600	3,000	4,200	3,500
Total working population	3,900	3,200	4,540	3,670

QUESTION   ANSWER

1. How many more self-employed workers were there in 2007 than in 2005?

    a.   5
    b.   10
    c.   20
    d.   30
    e.   40

QUESTION   ANSWER

2. What was the ratio of male to female workers for
   the total working population in 2005?
   a.   1.22:1
   b.   1.24:1
   c.   1.27:1
   d.   1.30:1
   e.   1.32:1

QUESTION   ANSWER

3. What percentage of the total working population
   in 2007 were female?
   a.   39%
   b.   41%
   c.   43%
   d.   45%
   e.   47%

QUESTION   ANSWER

4. By what percentage did the other-employed male
   workers increase from 2005 to 2007?
   a.   15.5%
   b.   16.2%
   c.   16.7%
   d.   17.1%
   e.   17.6%

QUESTION   ANSWER

5. By what percentage did the self-employed female
   workers decrease from 2005 to 2007?
   a.   6.0%
   b.   8.0%
   c.   10.0%
   d.   13.0%
   e.   15.0%

QUESTION   ANSWER

6. If the total working population in 2005 reflects
   a 25% increase since 2000, what was the total
   working population in 2000?
   a.   5,500
   b.   5,680
   c.   5,720
   d.   5,850
   e.   5,920

QUESTION   ANSWER

7. If in 2003 the total working population was 95%
   of what it was in 2005, and the ratio of other- to
   self-employed workers was 1.5:1, how many
   other-employed workers were there in 2003?
   a.   4,012
   b.   4,026
   c.   4,047
   d.   4,078
   e.   4,112

QUESTION   ANSWER

8. If the total working population increased by
   10% from 2007 to 2009, but the male to female
   ratio remained the same, how many male
   workers were there in 2009?
   a.   4,037
   b.   4,892
   c.   4,909
   d.   4,918
   e.   4,994

FIGURE 2: **Retail sales**

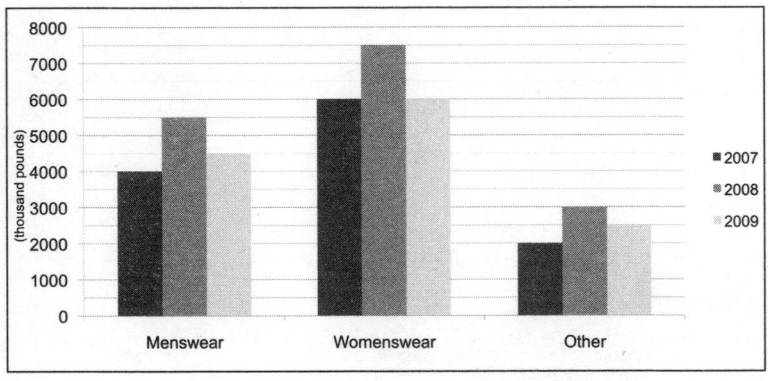

QUESTION ANSWER

9. What were the total sales made in 2008?
   a.   £13,000
   b.   £16,000
   c.   £12,000,000
   d.   £13,000,000
   e.   £16,000,000

QUESTION ANSWER

10. What were the average annual sales in menswear
   between 2007 and 2009?
   a.   £4,666,667
   b.   £4,667,000
   c.   £5,200,000
   d.   £5,500,000
   e.   £6,500,000

QUESTION ANSWER

11. If the womenswear sales increase from 2009 to
   2010 by 10%, what will the sales be in 2010?
   a.   £6,200,000
   b.   £6,400,000
   c.   £6,600,000
   d.   £6,800,000
   e.   £7,000,000

12. By what percentage did the sales in womenswear
increase from 2007 to 2008?

   a.   20%

   b.   25%

   c.   27%

   d.   30%

   e.   32%

13. By what percentage did the sales in 'other'
decrease from 2008 to 2009?

   a.   16.67%

   b.   17.00%

   c.   17.50%

   d.   18.33%

   e.   20.16%

14. If the average annual sales in womenswear
were £6,800,000 between 2007 and 2010, what
were the sales in womenswear in 2010?

   a.   £7,000,000

   b.   £7,500,000

   c.   £7,700,000

   d.   £8,000,000

   e.   £8,200,000

QUESTION   ANSWER

15. If the menswear sales increase from 2009 by 5%
annually, what will they be in 2012?
  a.   £4,725,000
  b.   £4,961,250
  c.   £5,175,000
  d.   £5,209,313
  e.   £5,469,778

QUESTION   ANSWER

16. What was the ratio of menswear to womenswear
to other, in 2007?
  a.   1.5:3:1
  b.   2:3:1
  c.   2.5:3:1
  d.   2:2.5:1
  e.   3:2:1

FIGURE 3: **Mobile phone subscriptions**

		Mobile phone subscriptions (thousands)	Mobile phone lines (thousands)	Population (thousands)
1995	Austria	300	3,670	7,660
	Malta	11	180	380
	Italy	3,820	22,640	54,500
	Estonia	25	345	1,230
	Finland	980	2,700	4,920
2003	Austria	7,200	3,270	8,150
	Malta	290	210	400
	Italy	55,680	26,100	57,810
	Estonia	1,000	440	1,280
	Finland	4,730	2,550	5,200

QUESTION   ANSWER

17. How many main phone lines were there in
    Malta and Estonia in 2003 (in thousands)?
    a.   525
    b.   640
    c.   650
    d.   1,290
    e.   2,990

QUESTION   ANSWER

18. On average, how many mobile phone
    subscriptions were there in 1995 among the
    five countries presented (in thousands)?
    a.   1,038
    b.   1,043
    c.   1,052
    d.   1,059
    e.   13,780

QUESTION   ANSWER

19. By what percentage did the population in Austria
    increase from 1995 to 2003?
    a.   5.68%
    b.   5.96%
    c.   6.06%
    d.   6.23%
    e.   6.54%

20. What was the ratio of mobile phone subscriptions to main phone lines in Finland in 1995?

a.   0.07:1

b.   0.17:1

c.   0.36:1

d.   1.86:1

e.   2.27:1

21. If the population in Italy increased from 2003 to 2009 by the same percentage as it increased from 1995 to 2003, what was the population in Italy in 2009 (in thousands)?

a.   59,244

b.   61,120

c.   61,259

d.   61,321

e.   62,057

22. If the population in Austria grew by 0.15% per year from 2003, what was the population in 2007 (in thousands)?

a.   8,199

b.   8,205

c.   8,211

d.   8,231

e.   8,240

QUESTION   ANSWER

23. If the mobile phone subscriptions increased by
2% in Finland and by 5% in Austria, from 2003
to 2004, what was the ratio of mobile phone
subscriptions of Finland to Austria in 2004?

    a.   0.632:1

    b.   0.638:1

    c.   0.645:1

    d.   0.657:1

    e.   0.660:1

QUESTION   ANSWER

24. If the number of main phone lines decreased by
2% in both Malta and Estonia from 2003 to 2008,
how many more main phone lines were there in
Estonia than in Malta in 2008 (in thousands)?

    a.   225

    b.   227

    c.   230

    d.   232

    e.   235

FIGURE 4: Employees by sector

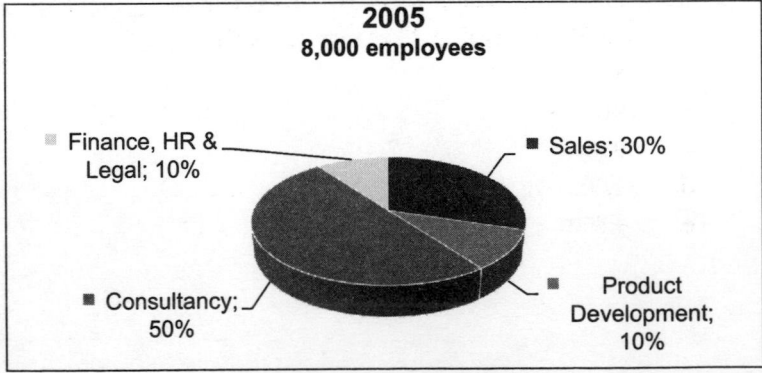

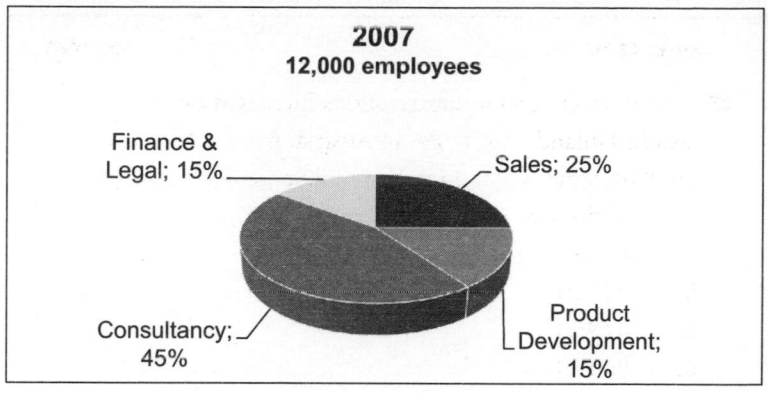

2007
12,000 employees

Finance & Legal; 15%

Sales; 25%

Consultancy; 45%

Product Development; 15%

QUESTION   ANSWER

25. How many employees were there in Sales in
    2007?
    a.   3,000
    b.   3,140
    c.   3,200
    d.   2,600
    e.   4,000

QUESTION   ANSWER

26. How many more employees were there in
    Finance, HR & Legal in 2007 than in 2005?
    a.   800
    b.   900
    c.   950
    d.   1,000
    e.   1,800

27. What was the ratio of consultants to sales staff
in 2005?

   a.   1.60:1

   b.   1.67:1

   c.   1.72:1

   d.   1.78:1

   e.   1.81:1

28. What was the ratio of consultants in 2005 to
consultants in 2007?

   a.   0.74:1

   b.   0.82:1

   c.   0.88:1

   d.   0.90:1

   e.   1.35:1

29. If the number of employees increased by 25%
from 2000 to 2005, how many staff were there
in 2000?

   a.   6,000

   b.   6,200

   c.   6,400

   d.   6,600

   e.   6,800

QUESTION   ANSWER

30. By what percentage did the number of employees in Product Development increase from 2005 to 2007?

    a.   25%

    b.   75%

    c.   125%

    d.   150%

    e.   225%

QUESTION   ANSWER

31. If the annual average between 2005 and 2007 was 9,500 employees, how many employees were there in 2006?

    a.   8,300

    b.   8,500

    c.   8,700

    d.   9,000

    e.   9,500

QUESTION   ANSWER

32. If the number of employees in Consultancy increased by 5% annually from 2007, how many employees were there in 2009?

    a.   5,954

    b.   6,424

    c.   6,528

    d.   6,530

    e.   6,534

Practice logical reasoning test

Instructions

This is a test assessing your ability to reason in the abstract. Each item in the test consists of a series of diagrams, which follow a logical sequence. Your task is to identify the missing diagram from the series (i.e. the box with the question mark), out of the response options given.

1.

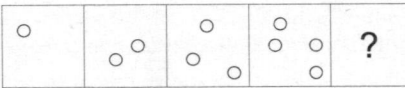

A.   B. ⊞ C. ⊞ D. ⊞ E. 8

2.

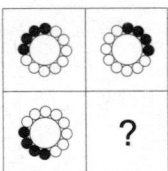

A.   B. C.   D.   E.

3.

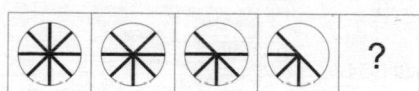

A.    B. 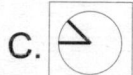   C. 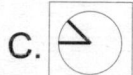   D.   E.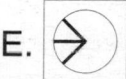

4.

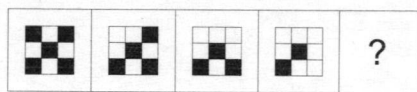

A. 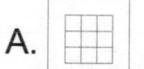   B.   C. 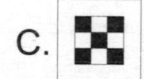   D.   E.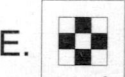

5.

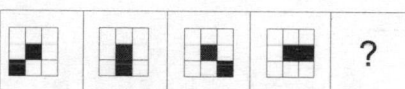

A. 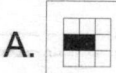   B.   C. 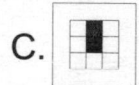   D.    E.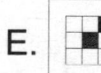

6.

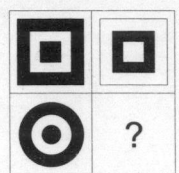

7.

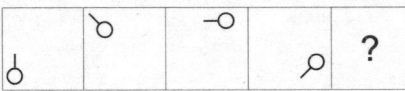

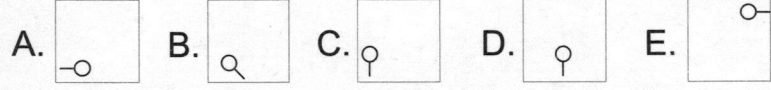

8.

9.

A.  B. 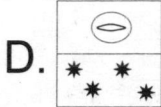 C. D.

10.

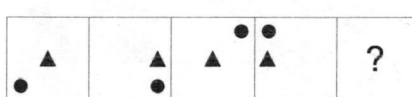

A. 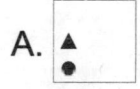 B. C. 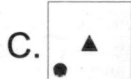 D. E.

11.

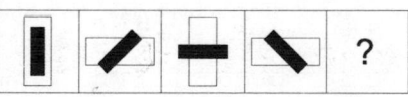

A. 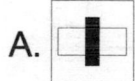 B.  C. 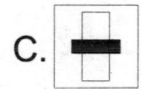 D. 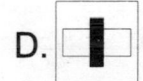 E.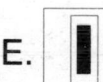

12.

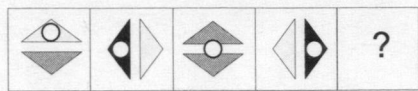

A.  B.  C. 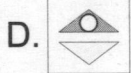 D.  E.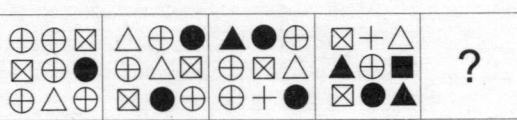

13.

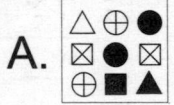

A. 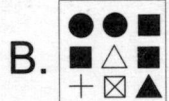 B. 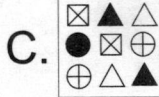 C. 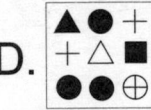 D.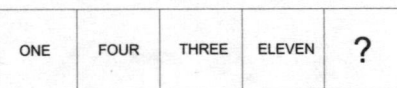

14.

ONE	FOUR	THREE	ELEVEN	?

A. ZERO   B. TWO   C. SIXTEEN   D. TWENTY   E. FIVE

15.

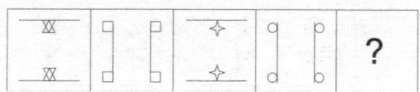

A. 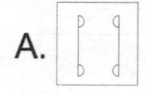   B.   C.   D.   E.

16.

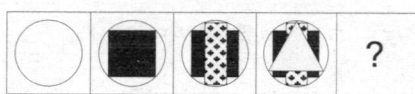

A.    B.   C.    D.   E.

17.

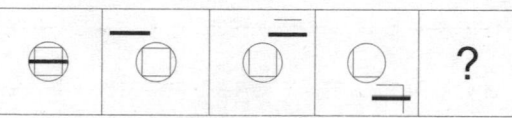

A. 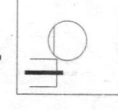   B. 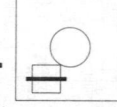   C. 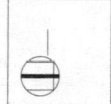   D.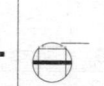

18.

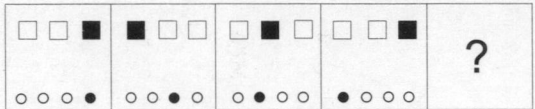

A. 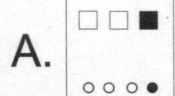　B. 　C. 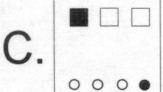　D.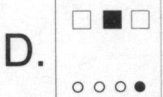

19.

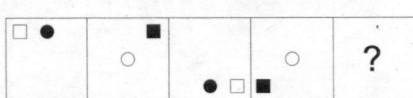

A. 　B. 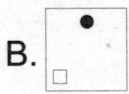　C. 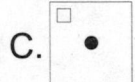　D. 　E.

20.

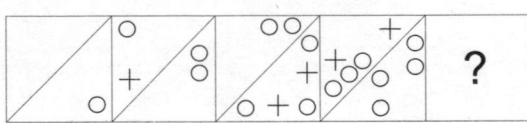

A. 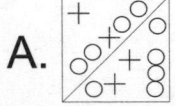　B. 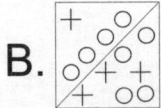　C. 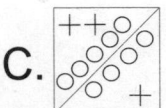　D.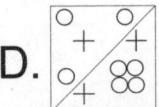

21.

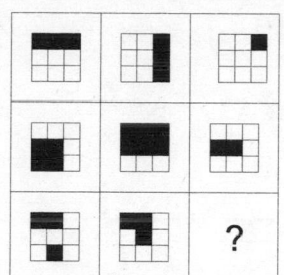

A. 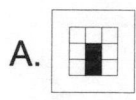 B. 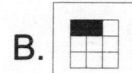 C. 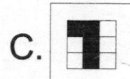 D. 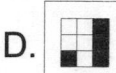 E.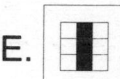

22.

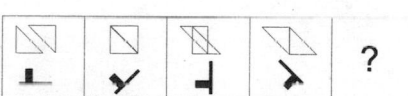

A.  B. 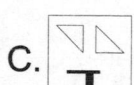 C. 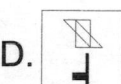 D. 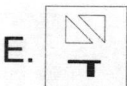 E.

23.

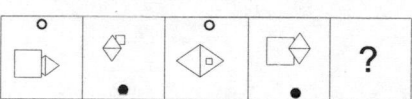

A. 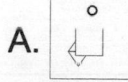 B. 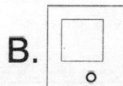 C. 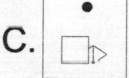 D. 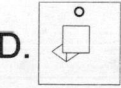 E.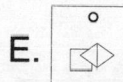

24.

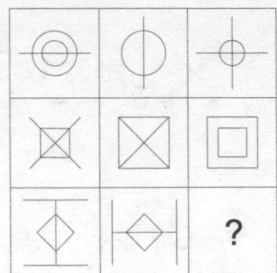

A. 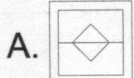 B. 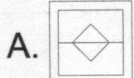 C. 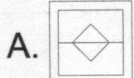 D. 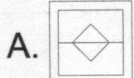 E.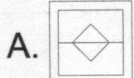

25.

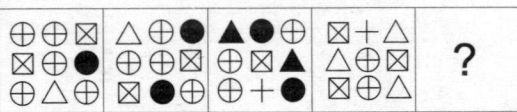

A.  B.  C.  D.

26.

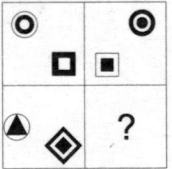

A.  B.  C.  D.  E.

27.

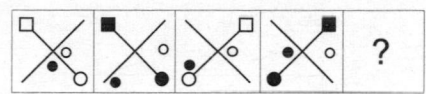

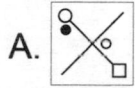

28.

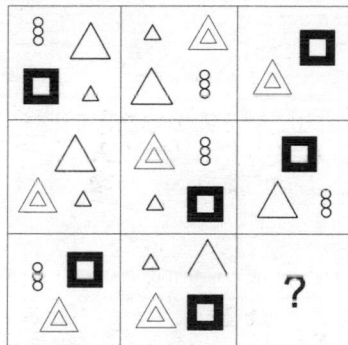

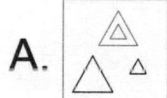

29.

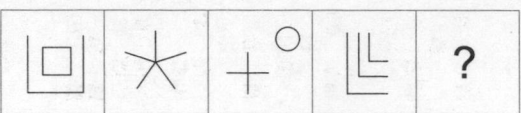

A.   B. C.   D.

30.

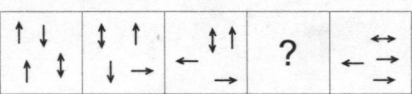

A. B. C. D. E.

31.

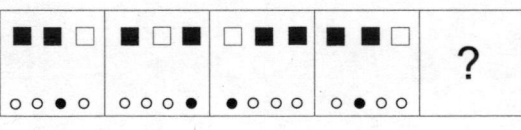

A. 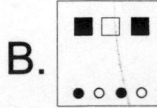 B. C. D.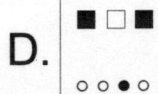

32.

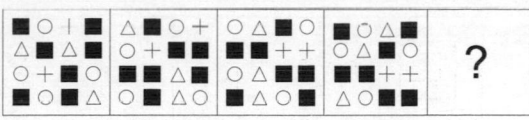

A. 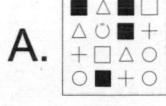 B. 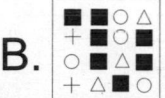 C. 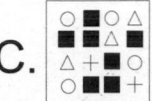 D.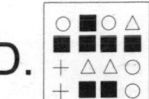

Further reading

If you would like to practise some more numerical and logical tests, there are several websites where you can take practice tests online. This will be especially useful if you have been asked to sit an online test, as it will give you the opportunity to familiarise yourself with the test conditions.

If possible, try to take some timed practice tests, so that you can also evaluate how you are doing with the added pressure of time.

Here are some useful websites:

http://www.shl.com/tryatest/takeatest/default.aspx
http://practicetests.cubiks.com
http://www.kenexa.com/Solutions/Assessment/Abilities.aspx
http://www.kent.ac.uk/careers/tests/mathstest.htm
http://www.assessmentday.co.uk
http://www.psychometric-success.com/downloads/download-practice-tests.htm
http://www.morrisby.com/content/candidates-support/faqs/sample-morrisby-profile-questions/index.htm
http://www.mensa.org/workout2.php
http://www.savilleconsulting.com/products/aptitude_preparation guides.aspx
http://www.previsor.co.uk

As some of the links may have changed since the book was published, you could also do an internet search using key words. If you do that, try to select tests that are published by an academic or professional body, to ensure that the tests you are practising on are reliable and valid.

If you are interested in learning more about numerical tests, or would like to find out about other types of psychometric tests, you could read:

Joanna Moutafi and Ian Newcombe, *Perfect Numerical Test Results* (London, 2007)

Joanna Moutafi and Ian Newcombe, *Perfect Psychometric Test Results* (London, 2007)

You can also read about personality profiles in:

Helen Baron, *Perfect Personality Profiles* (London, 2007)

If you want to improve your performance in other parts of the selection process, for example if you want to improve your CV or improve your performance at interviews, you could also read:

Max Eggert, *Perfect CV* (London, 2003)

Max Eggert, *Perfect Interview* (London, 2003)

Max Eggert, *Perfect Answers to Interview Questions* (London, 2005)

And finally, if you want to learn more about the fairness of psychometric tests and what reliability and validity mean, you can read:

Edward G. Carmines and Richard A. Zeller, *Reliability and Validity Assessment (Quantitative Applications in the Social Sciences)* (Newbury Park, 1979)

Final comment

Congratulations! You have successfully reached the end of the book. You have completed all the steps that will help you prepare for your battle and achieve your goal: to get the job you are after.

By now, you should be familiar with two types of psychometric tests – numerical reasoning and logical reasoning. You know what types of questions you are likely to face in each one, and how to find the answers accurately and quickly. You have also learned the best way to prepare yourself for these tests, both mentally and physically. Finally, you have been given nearly 300 examples to practise before the big day of the test. It is up to you now. Follow our advice and prepare to perform at your best.

We wish you all the best for a successful career!

Answers to practice questions

Step 3: Master your opponent – numerical tests

Question	Answer		Question	Answer
Addition			**Division**	
1.	61		16.	8
2.	23.21		17.	4
3.	102		18.	9
4.	620		19.	7
5.	1.1		20.	18
Subtraction			**Rounding off numbers**	
6.	82		21.	13.5
7.	59		22.	1.5
8.	0.11		23.	1.5
9.	417		24.	2.3
10.	−24		25.	1.0
			26.	24.84
			27.	145.45
Multiplication			28.	4.30
11.	56		29.	0.82
12.	54		30.	2.00
13.	60			
14.	56		**Transforming numbers and percentages**	
15.	39			
			31.	17%
			32.	24.2%

Question	Answer
33.	410%
34.	102%
35.	10,000
36.	2.3%
37.	198%
38.	0.290
39.	0.355
40.	0.421
41.	0.008
42.	0.003
43.	1.000
44.	1.200
45.	10.000
46.	1.068
47.	50%
48.	20%
49.	25%
50.	200%
51.	80%
52.	50%
53.	500%
54.	75%
55.	25%

Averages

Question	Answer
56.	d
57.	b
58.	c
59.	e
60.	d
61.	c

Question	Answer
62.	b
63.	a
64.	b
65.	b
66.	d
67.	d
68.	e
69.	a
70.	b
71.	c
72.	d
73.	e

Percentages

Question	Answer
74.	c
75.	d
76.	d
77.	d
78.	e
79.	d
80.	a
81.	b
82.	d
83.	b
84.	a
85.	a
86.	e
87.	b
88.	a
89.	c
90.	c

Question	Answer		Question	Answer
91.	c		**Ratios**	
92.	e		122.	a
93.	c		123.	b
94.	d		124.	e
95.	a		125.	d
96.	d		126.	a
97.	c		127.	a
98.	e		128.	d
99.	d		129.	a
100.	d		130.	c
101.	a		131.	e
102.	c		132.	d
103.	b		133.	a
104.	a		134.	b
105.	d		135.	d
106.	e		136.	c
107.	d		137.	b
108.	c		138.	d
109.	e		139.	a
110.	b		140.	e
111.	b		141.	e
112.	d		142.	b
113.	a		143.	b
114.	c		144.	c
115.	e		145.	d
116.	d		146.	e
117.	d			
118.	a		**Combining averages, ratios and**	
119.	e		**percentages**	
120.	c		147.	c
121.	d		148.	d

Question	Answer
149.	a
150.	b
151.	e
152.	c
153.	d
154.	c
155.	c
156.	c
157.	a
158.	b
159.	e
160.	b
161.	c

Unit conversions

Question	Answer
162.	e
163.	a
164.	c
165.	d
166.	b
167.	e
168.	c
169.	a
170.	a
171.	d
172.	b
173.	c
174.	e
175.	e
176.	a

Sequences

Question	Answer
177.	d
178.	b
179.	c
180.	d
181.	d
182.	c
183.	a
184.	e
185.	d
186.	e
187.	b
188.	c
189.	a
190.	e
191.	b
192.	d
193.	c
194.	d
195.	c
196.	d
197.	b
198.	e
199.	d
200.	c
201.	b
202.	d
203.	e

Step 5: Face your opponent - practice tests

Numerical test

Question	Answer
Figure 1	
1.	b
2.	a
3.	d
4.	c
5.	e
6.	b
7.	c
8.	e
Figure 2	
9.	e
10.	a
11.	c
12.	b
13.	a
14.	c
15.	d
16.	b
Figure 3	
17.	c
18.	b

Question	Answer
19.	e
20.	c
21.	d
22.	a
23.	b
24.	a
Figure 4	
25.	a
26.	d
27.	b
28.	a
29.	c
30.	c
31.	b
32.	a

Logical reasoning test

Question	Answer	Question	Answer
1.	c	17.	a
2.	b	18.	c
3.	a	19.	a
4.	b	20.	a
5.	e	21.	b
6.	a	22.	a
7.	c	23.	d
8.	e	24.	d
9.	d	25.	a
10.	c	26.	d
11.	e	27.	a
12.	b	28.	b
13.	d	29.	c
14.	c	30.	b
15.	b	31.	d
16.	d	32.	c

ALSO AVAILABLE IN RANDOM HOUSE BOOKS

Perfect Answers to Interview Questions

Max Eggert

All you need to get it right first time

- Are determined to succeed in your job search?
- Do you want to make sure you stand out from the competition?
- Do you want to find out what interviewers really want to hear?

Perfect Answers to Interview Questions is essential reading for anyone who's applying for jobs. Written by a leading HR professional with years of experience in the field, it explains the sorts of questions most frequently asked, gives practical advice about how to show yourself in your best light, and provides real-life examples to help you practise at home. Whether you're a graduate looking to take the first step on the career ladder, or you're planning an all-important job change, *Perfect Answers to Interview Questions* will give you the edge.

BOOKS

Perfect Brain Training

Philip Carter

All you need to boost your brainpower

- Do you sometimes find yourself getting confused or forgetting things?
- Are you worried that your thinking is not as sharp as it used to be?
- Do you want a simple way to improve your brainpower?

Perfect Brain Training is essential reading for anyone who wants to improve their mental agility. Containing a series of fun interactive workouts, it helps you develop every aspect of your thinking skills, from logical deduction and creative problem solving to memory power and verbal dexterity. With advice to help you put together an effective training programme and tests so that you can track your progress, *Perfect Brain Training* has everything you need to make the most of your potential.

BOOKS

Perfect Confidence

Jan Ferguson

All you need to get it right first time

- Do you find it hard to stay calm under pressure?
- Are you worried that you don't always stand up for yourself?
- Do you want some straightforward advice on overcoming insecurities?

Perfect Confidence is the ideal companion for anyone who wants to boost their self-esteem. Covering everything from communicating clearly to handling conflict, it explains exactly why confidence matters and equips you with the skills you need to become more assertive. Whether you need to get ahead in the workplace or learn how to balance the demands of friends and family, *Perfect Confidence* has all you need to meet challenges head on.

BOOKS

Perfect CV

Max Eggert

All you need to get it right first time

- Are you determined to succeed in your job search?
- Do you need guidance on how to make a great first impression?
- Do you want to make sure your CV stands out?

Bestselling *Perfect CV* is essential reading for anyone who's applying for jobs. Written by a leading HR professional with years of experience, it explains what recruiters are looking for, gives practical advice about how to show yourself in your best light, and provides real-life examples to help you improve your CV. Whether you're a graduate looking to take the first step on the career ladder, or you're planning an all-important job change, *Perfect CV* will help you stand out from the competition.

BOOKS

Perfect Interview

Max Eggert

All you need to get it right first time

- Are you determined to succeed in your job search?
- Do you want to make sure you have the edge on the other candidates?
- Do you want to find out what interviewers are really looking for?

Perfect Interview is an invaluable guide for anyone who's applying for jobs. Written by a leading HR professional with years of experience in the field, it explains how interviews are constructed, gives practical advice about how to show yourself in your best light, and provides real-life examples to help you practise at home. Whether you're a graduate looking to take the first step on the career ladder, or you're planning an all-important job change, *Perfect Interview* will help you stand out from the competition.

BOOKS

Perfect Memory Training

Fiona McPherson

All you need to get it right every time

- Do you sometimes find it hard to remember names, dates or where you left your keys?
- Would you like to find out how your memory works?
- Do you want to learn some simple skills to help you stop forgetting things?

Perfect Memory Training is essential reading for anyone who wants to strengthen their powers of recall. Written by Dr Fiona McPherson, a psychologist with years of experience in the field, it explains how memories are created and stored, sets out a range of techniques to help you improve these processes, and provides exercises to help you track your progress. Whether you want to get better at remembering names, faces, lists or pieces of general knowledge, *Perfect Memory Training* has everything you need to boost your mental ability.

BOOKS

Perfect Numerical Test Results

Joanna Moutafi and Ian Newcombe

All you need to get it right first time

- Have you been asked to sit a numerical reasoning test?
- Do you want guidance on the sorts of questions you'll be asked?
- Do you want to make sure you perform to the best of your abilities?

Perfect Numerical Test Results is an invaluable guide for anyone who wants to secure their ideal job. Written by a team from Kenexa, one of the UK's leading compilers of psychometric tests, it explains how numerical tests work, gives helpful pointers on how to get ready, and provides professionally constructed sample questions for you to try out at home. It also contains an in-depth section on online testing – the route that more and more recruiters are choosing to take. Whether you're a graduate looking to take the first step on the career ladder, or you're planning an all-important job change, *Perfect Numerical Test Results* has everything you need to make sure you stand out from the competition.

BOOKS

Perfect Persuasion

Richard Storey

All you need to get it right first time

- Do you have difficulty getting people to agree with you?
- Are you worried that your voice isn't being heard?
- Do you need pointers on ways to win over friends and colleagues?

Perfect Persuasion is essential reading for anyone who wants to improve their powers of influence. Written by Richard Storey, an expert with years of experience in the field, it explains how to identify other people's motivations, gives practical advice about staying calm when faced with resistance, and takes you through every skill you need to win people over to your point of view. Whether you need to influence colleagues at work or would like to make some changes in your personal life, *Perfect Persuasion* has everything you need to make sure you get your point across effectively.

BOOKS

Perfect Presentations

Andrew Leigh and Michael Maynard

All you need to get it right every time

- Have you been asked to give a presentation?
- Would you like some guidance on the best way to deliver your material?
- Do you want to make sure you get your message across effectively?

Perfect Presentations is an invaluable guide for anyone preparing to speak in public. Written by two professional trainers with years of experience in the field, it explains how to plan and structure talks, offers tips on conquering nerves, and gives suggestions for the most effective and inspiring way to deliver your material. Whether you're taking your first steps on the career ladder and want some pointers, or you're a seasoned professional looking to refine your presenting technique, *Perfect Presentations* has all you need to make sure you come across brilliantly.

BOOKS

Perfect Personality Profiles

Helen Baron

All you need to make a great impression

- Have you been asked to complete a personality questionnaire?
- Do you need guidance on the sorts of questions you'll be asked?
- Do you want to make sure you show yourself in your best light?

Perfect Personality Profiles is essential reading for anyone who needs to find out more about psychometric profiling. Including everything from helpful pointers on how to get ready to a thorough breakdown of the personality traits that questionnaires examine, it walks you through every aspect of personality profiles. Whether you're a graduate looking to take the first step on the career ladder, or you're planning an all-important job change, *Perfect Personality Profiles* has everything you need to make sure you stand out from the competition.

BOOKS

Perfect Psychometric Test Results

Joanna Moutafi and Ian Newcombe

All you need to get it right first time

- Have you been asked to sit a psychometric test?
- Do you want guidance on the sorts of questions you'll be asked?
- Do you want to make sure you perform to the best of your abilities?

Perfect Psychometric Test Results is the ideal guide for anyone who wants to secure their ideal job. Written by a team from Kenexa, one of the UK's leading compilers of psychometric tests, it explains how each test works, gives helpful pointers on how to get ready, and provides professionally constructed sample questions for you to try out at home. It also contains an in-depth section on online testing – the route that more and more recruiters are choosing to take. Whether you're a graduate looking to take the first step on the career ladder, or you're planning an all-important job change, Perfect Psychometric Test Results has everything you need to make sure you stand out from the competition.

BOOKS

Order more titles in the *Perfect* series
from your local bookshop, or have them delivered
direct to your door by Bookpost.

☐ Perfect Answers to Interview Questions	Max Eggert	9781905211722	£7.99
☐ Perfect Brain Training	Philip Carter	9781847945549	£6.99
☐ Perfect Confidence	Jan Ferguson	9781847945693	£7.99
☐ Perfect CV	Max Eggert	9781905211739	£7.99
☐ Perfect Interview	Max Eggert	9781905211746	£7.99
☐ Perfect Memory Training	Fiona McPherson	9781847945365	£7.99
☐ Perfect Numerical Test Results	Joanna Moutafi and Ian Newcombe	9781905211678	£7.99
☐ Perfect Persuasion	Richard Storey	9781847945594	£7.99
☐ Perfect Presentations	Andrew Leigh and Michael Maynard	9781847945518	£6.99
☐ Perfect Personality Profiles	Helen Baron	9781905211821	£7.99
☐ Perfect Psychometric Test Results	Joanna Moutafi and Ian Newcombe	9781905211678	£7.99

Free post and packing
Overseas customers allow £2 per paperback
Phone: 01624 677237

Post: Random House Books
c/o Bookpost, PO Bow 29, Douglas, Isle of Man IM99 1BQ

Fax: 01624 670 923

email: bookshop@enterprise.net

Cheques (payable to Bookpost) and credit cards accepted

Prices and availability subject to change without notice.
Allow 28 days for delivery.
When placing your order, please state if you do not
wish to receive any additional information.

www.rbooks.co.uk

BOOKS